Taylor Downing is a television producer and writer. He was educated at Latymer Upper and Christ's College, Cambridge where he took a Double First in History. He worked at the Imperial War Museum and for several years has run Flashback Television, an independent production company, where he has produced more than two hundred documentaries, including many award-winning historical films. He has also written several history books. Taylor Downing is married and lives in London and East Devon.

Churchill's War Lab

CODE-BREAKERS, BOFFINS AND INNOVATORS:

THE MAVERICKS CHURCHILL LED TO VICTORY

Taylor Downing

ABACUS

First published in Great Britain in 2010 by Little, Brown
This paperback edition published in 2011 by Abacus

A CIP catalogue record for this book
is available from the British Library.

ISBN 978-0-349-12249-6

Typeset in Palatino by M Rules
Printed and bound in Great Britain by
Clays Ltd, St Ives plc

Papers used by Abacus are from well-managed forests
and other responsible sources.

MIX
Paper from
responsible sources
FSC® C104740

Abacus
An imprint of
Little, Brown Book Group
100 Victoria Embankment
London EC4Y 0DY

An Hachette UK Company
www.hachette.co.uk

www.littlebrown.co.uk

For Anne
With thanks and for support

And for my Father
Who started my interest in all this

Contents

Introduction

A summer's day, 1886. The sunlight falls brightly across the nursery floor of a rather grand house in a smart street in London. Two boys are playing with their toy soldiers on the floor. The younger of the two plays in a desultory way. His heart is not in the game. His elder brother, on the other hand, is the very picture of concentration and seriousness. He moves his soldiers with utmost care and attention. He commands an army of nearly fifteen hundred troops. They are all perfectly painted in the colours of the British Army. Different regiments stand out clearly in their smart field uniforms. And they are properly organised into an infantry division with a cavalry brigade on the flank. There are artillery pieces as well: eighteen field artillery guns and a few heavy pieces for assaulting solid fortresses. The older boy has arranged his troops into a perfect formation of attack.

This afternoon the two boys' father comes to pay a visit. He is a very important man, a leading politician. Indeed, he has just been appointed Chancellor of the Exchequer in the Conservative government. Hence, he is always very busy and has little time for his sons; nevertheless, they adore him. On this occasion the father spends a full twenty minutes in the nursery, studying the impressive scene of an army ready to launch an assault upon its foe. At the end of his inspection, he asks his eldest son if he would like to go into the army when he grows up. Winston Churchill replies at once: yes, he would love to. It would be splendid to command a real army. After this day, the young Winston's education will be focused on getting him into the Royal Military College at Sandhurst, where he will learn the technical details of the profession of arms. The boy's obsession and fascination with his toy soldiers had, as he later wrote, 'turned the current' of his life. Winston Churchill would grow up to be a soldier.[1]

August 1898. The boy is now a distinguished young officer in one of the elite cavalry regiments of the British Army, the 4th Hussars. The only trouble is that the 4th Hussars are stationed in southern India. And there is no action in southern India. Seeking out the thrill of military combat, the twenty-three-year-old subaltern arranges a transfer to the army led by Sir Herbert Kitchener that is mounting an expedition into the Sudan. Churchill is temporarily enlisted with the 21st Lancers, who are part of this vast Anglo-Egyptian army of twenty-five thousand men that is slowly travelling down the west bank of the river Nile, teasing out an engagement with the Muslim Dervish army near Khartoum. He is enthralled by the magnificent sight of this advancing army with its five brigades, each of three or four infantry battalions, marching in open columns across the sandy desert, along with its artillery and transport, supported by a flotilla of grey gunboats sailing down the Nile. The 21st Lancers patrol the flank and scout ahead for the enemy.

On 1 September the Dervish army is sighted, fifty thousand strong, assembling in huge phalanxes. At dawn on the following day, battle ensues. Churchill watches the early stages of the battle from the top of a ridge while passing reports back to his commanding officers. As the sun slowly comes up he is exhilarated by the experience: 'Talk of Fun! Where will you beat this! On horseback, at daybreak, within shot of an advancing army, seeing everything and corresponding direct with Headquarters.'

The battle that follows sees a modern, well-equipped nineteenth-century army engage with a massed force of local tribesmen. The shells and bullets of British howitzers, Maxim guns and carbine rifles tear into the Dervish soldiers, creating huge, deadly swaths in their ranks. The result is a foregone conclusion. Within hours, the Anglo-Egyptian army wins a tremendous victory. But the Dervishes are tough, well-trained and highly motivated soldiers. Later that morning, as the 21st Lancers escort the infantry towards the enemy capital, they come under fire from their right flank. The trumpets and bugles sound the order first to 'Trot', then to 'Wheel Right', then to 'Charge', and the Lancers carry out a manoeuvre they have long trained for, a cavalry charge at full gallop and in close order against an enemy line. Churchill, leading his troop of some twenty-five men, rides right through the line of Dervish defenders, riflemen and spear carriers. But the line holds. Turning around on the other side, Churchill finds himself isolated and surrounded by ferocious Dervishes. He shoots at least three men at close range, rallies his troop and they regroup. Then they open rapid fire on the enemy, who cannot survive this enfilade. In twenty minutes the action is over. The Dervishes withdraw along a wadi, a sunken riverbed, carrying their wounded with them. Churchill is unscathed but counts his losses. His troop has done well, but his regiment of 310 officers and men has lost 5 officers and 65 men killed and wounded in just a few minutes.

Churchill has taken part in the last great cavalry charge in history, at the Battle of Omdurman. Noble and magnificent though it might have seemed, charging across the desert to the echo of hoofs and the clatter of reins, the officers with their swords raised, in reality the cavalry charge was a futile act on the fringes of the battle. It contributed nothing to the enemy's defeat and only inflicted losses of nearly one man in four on the 21st Lancers – far higher than those suffered by any other British unit that day. However, the event was covered with glory and the Lancers won three Victoria Crosses that morning. Churchill later described the battle in a way that encapsulates how many people saw wars on the fringes of the empire in the late nineteenth century: as good sport. He wrote, 'This kind of war was full of fascinating thrills . . . No one expected to be killed . . . [T]o the great mass of those who took part in the little wars of Britain in those vanished light-hearted days, this was only a sporting element in a splendid game.'[2]

Afternoon, 1 June 1944. A small group gathers in the Map Room located at the heart of the underground War Rooms, a top-secret bunker constructed for government leaders in what is called the Downing Street Annexe. The group discusses plans for the D-Day invasion, which everyone knows is now only a few days away. The King is at this briefing, along with his private secretary, Sir Alan Lascelles. The Prime Minister is also present, as well as a few top military figures. At the meeting the Prime Minister presses his desire to be present at the landings on the Normandy coast. He wants to be on board HMS *Belfast*, the flagship of the naval commander of the British fleet at this historic moment. It is agreed that there are risks associated with this: the ship could be bombed, hit by shells or struck by a mine. The largest amphibious landings in history will be taking place only a few thousand yards away.

The Prime Minister asks the King if he would like to be present also, and to lead his troops into battle, like monarchs in

olden days. Lascelles is utterly horrified at the idea, feeling that
neither the King nor the Prime Minister should put his life at
such appalling risk. His face grows longer and longer. Eventually he speaks up and asks how the King would feel if he had
to find a new prime minister during the middle of the D-Day
landings. The Prime Minister dismisses this possibility, but the
King argues that it is foolish of him knowingly to put himself
in the face of such danger in what is a 'joy ride'. The Prime
Minister replies that during the course of the war he has flown
to the United States and the Middle East, and has crossed the
Atlantic many times: sometimes he needs to take risks in order
to carry out his duties.

The King leaves the meeting and returns to Windsor. On the
following day, he resolves to instruct the Prime Minister not to
witness the D-Day landings in person. He writes him a letter in
which he claims: 'you will see very little, you will run a considerable risk, you will be inaccessible at a critical time when
vital decisions might have to be taken; and however unobtru-
sive you may be, your mere presence on board is bound to be
a heavy additional responsibility to the Admiral & Captain'.
Later that day, the Prime Minister relents and reluctantly
accepts the instruction of his sovereign not to travel to the battle-
front.[3]

This story is again so typical of Winston Churchill. His generals and admirals are about to launch the long-awaited invasion
of Europe. It has taken years to prepare for this moment. The
assault on Fortress Europe will be one of the decisive moments
of the Second World War, and one of its fiercest battles. And
Churchill wants to be at the centre of it. He wants to witness the
action. He wants to see the dawn bombardment, observe the
landings, maybe even set foot on the beaches. He argues that
leaders of men at times of war sometimes need 'the refreshment
of adventure' and that his 'personal interest' is 'stimulated by

direct contact' with events. Although he offers the King the chance of being there too, in truth Churchill wants to lead the troops into battle himself. Just as he did at Omdurman. Just as he did when he laid out his toy soldiers as a boy.

Man and boy, soldiering and military history were part of Churchill's make-up, embedded in his DNA. For five years, from 1940 to 1945, he would oversee an extraordinary out-pouring of radical new ideas and fabulous new inventions in a sequence of events rich with brilliant but wacky boffins, remarkable mavericks and frustrated war chiefs. This would be Churchill's War Lab.

War is the mother of invention. A cliché, but true.

And no war generated more incredible ideas, more technical advances and more scientific leaps than the Second World War.

In the cauldron of ideas that simmered throughout that conflict, new inventions ranged from jet engines to roll-on/roll-off ferries, from flying wings to floating tanks, from minia-ture radios to guided missiles. Winston Churchill immersed himself in the work of his engineers and inventors, his soldiers, sailors and airmen, imprinting his own personality on the machines that were created in his name. Like no other British prime minister at a time of war, Churchill relished military debate and immersed himself in the work of the code-breakers and scientific mavericks who were needed to get the best out of Britain's sometimes low-key war effort. As a result of his encouragement, these men and women would eventually have a real impact on the outcome of the war.

The Second World War was fought as much by scientists, or 'boffins', as they were often known, as by soldiers, sailors and airmen. New ways of thinking, of approaching and assessing a question in the science known as Operational Research were applied to challenges facing the RAF, the British Army and the

Royal Navy. Operational Research applied scientific method to solving problems ranging from finding the optimum setting for depth charges, to the search pattern aircraft should follow when hunting U-boats, to the most effective way of siting and firing anti-aircraft guns. A tiny device only a few centimetres long and invented at Birmingham University, the cavity magnetron, opened up revolutionary new possibilities for short-wave radar. This could be used to guide bombers to their targets and to track the conning tower of a U-boat from dozens of miles away. In many ways the cavity magnetron was a war-winning invention; certainly it was hailed as such by the Americans when it was first shown off in Washington. And, of course, harnessing the power of the atom in the massive Manhattan Project, which employed some 120,000 scientists and workers, was literally a war-winning discovery. Science would also help to crack the codes used in top-level enemy communications, and would help to guide shells towards aircraft in the skies. One leading scientist said during the war that 'there is hardly a phase of the national life with which scientists are not associated' and that you could 'hardly walk in any direction in this war without tumbling over a scientist'.[4]

Churchill took a keen interest in the application of science to the technology of war. It is the underlying contention of this book that his encouragement of science and of new ways of approaching military challenges was at the core of Britain's final victory in the long struggle of the Second World War. At one point the head of Bomber Command, Sir Arthur 'Bomber' Harris, remonstrated with Churchill about a new approach the Prime Minister was championing.

'Are we fighting this war with weapons or slide rules?' Harris asked.

Churchill replied, 'That's a good idea; let's try the slide rule for a change.'[5]

Churchill was a dynamo who generated energy (and heat) at the heart of government. Often his senior officers resented his interference. General 'Pug' Ismay repeatedly had to mediate in rows between the Prime Minister and his service chiefs. One intimate noted that, without this control on his actions, 'Winston would have been a Caligula or worse, and quite properly [would have] had his throat cut.' His involvement was not restricted to matters of grand strategy: although he always had strong views on these, he would also involve himself in detailed tactical battlefield questions. Within a few weeks of taking over at 10 Downing Street, a visitor was astonished to hear Churchill having a phone call with a local field commander, arguing whether the 'brigadier . . . at Boulogne nearly 100 miles away was doing the right thing in resisting the Germans at one end of a quay or the other'.[6]

Churchill could be petulant and sometimes even childish. He constantly felt frustrated by what he perceived as the lack of drive in the military leaders who reported to him. He thought it was his job to bring vim and vigour to their deliberations and new ideas to their thinking. Once he remarked that taking an admiral out and having him shot would do a great deal 'to encourage the others'. He said that one of his leading generals was more suited to running a golf club than an army of fighting men. At one point, his wife Clementine reluctantly wrote to him that 'there is a danger of your being generally disliked by your colleagues and subordinates because of your rough, sarcastic and over-bearing manner'. But the pace of work quickened when Churchill was around. And most of those forced to work with him accepted that there were more pluses than minuses in his leadership.

Churchill became Prime Minister on the evening of 10 May 1940. Earlier that same day Hitler had launched his armies against Holland, Belgium and France. That evening, as a mighty battle raged on continental Europe, Churchill might have been justified in feeling overawed, even overwhelmed, by the role he had just taken up. Instead, he sensed that his 'whole life had been a preparation for this hour and this trial', and felt exhilarated that he was, in his famous phrase, 'walking with destiny'.[7]

The first two chapters of this book look at the key elements of Churchill's early life that prepared him for leadership in May 1940. He served in several regiments of the British Army both as a young man and in the trenches in 1915–16. He had regularly come under fire in combat. As a politician, he was President of the Board of Trade, Home Secretary and Chancellor of the Exchequer, First Lord of the Admiralty at the beginning of both world wars, and Minister of Munitions at the end of the first. During his so-called 'wilderness years' in the 1930s he honed his ideas about history and what he called England's 'special destiny'. All of this helped to shape the man who became Prime Minister at the age of sixty-five in Britain's hour of crisis.

Chapters 3 and 4 look in detail at his leadership during the critical year when Britain stood largely alone, until first the Soviet Union and then the United States turned a European conflict into a world war. The next four chapters step out of the chronology to look thematically at Churchill's relationships with the scientists who played leading roles in the war, alongside his generals, his admirals and his air marshals. As we shall see, these scientists who advised him along with the military chiefs were core members of his War Lab. Chapter 9 picks up the chronology of the war at the beginning of 1943, as Churchill returns from the Casablanca Conference with President Roosevelt, and at the events that lead up to Operation Overlord, the invasion of Northern Europe. Chapter 10 looks at Churchill's role during the final year of the war, while the concluding chapter offers a brief assessment of his wartime leadership.

Every new book on Churchill has to justify its existence in the crowded market place of Churchilliana. This one has emerged out of years of making television programmes about the Second World War and meeting some of the key participants in that war. It comes from a realisation that, although the Allies certainly did not have a monopoly on good science and technology (far from it), the application of this scientific approach under Churchill's encouragement contributed significantly to their ultimate victory.

My father was a government defence scientist who had been recruited into the RAF straight out of university in 1942. He certainly regarded himself as a boffin, one of the 'backroom boys'. Looking back now, I suppose he left me with a lasting interest in the relationship between science, technology and war. So, *Churchill's War Lab* presents a new take on the remarkable years of Churchill's war leadership. It is partly about the technology of war, partly about how Churchill was forged into the sort of war leader he was, and partly about how he inspired the mavericks and innovators to go out and influence the course of the Second World War. Many of the characters who appear in this book deserve books of their own. But Churchill himself rightly occupies centre stage throughout.

Taylor Downing
October 2009

1

Preparation: The Army and the Navy

When Winston Churchill became Prime Minister in May 1940, to lead the British nation in a war for its survival, he knew more about military affairs and soldiering than any other wartime British premier. Much more than William Pitt the Younger and Spencer Perceval, who led Britain at the time of the Napoleonic Wars, when the country was threatened with invasion by Napoleon. More than Lord Palmerston, who was brought in to lead the government when the Crimean War broke out in 1854. More than Asquith and Lloyd George, who led Britain through the appalling sacrifices of the Great War. And certainly much more than recent British prime ministers who have taken the nation to war – Margaret Thatcher in the Falklands in 1982, John Major in the First Gulf War in 1991, Tony Blair in Iraq in 2003 and Gordon Brown, who inherited the war in Afghanistan. Churchill had trained as a soldier, had served in several regiments of the British Army, had considerable experience of coming under fire, had been captured and had escaped, had led men in battle, and had fought in the trenches in the First World

War. He had studied the fighting of wars and had written famous military histories. He had been in overall command of the Royal Navy in an era when Britannia unarguably ruled the waves. He had led a life that had been imbued with military matters. And he had loved it. It should not be surprising, then, that when he came to lead the nation in war he would run his government in a different way to any other war leader in history. This is that story. But first, it is necessary to see how his previous life was, as he later wrote, 'preparation for this hour and this trial'.

Winston Leonard Spencer Churchill was born on 30 November 1874 into the fringes of one of Britain's greatest aristocratic families. His ancestor, John Churchill, had led an army against France in the War of the Spanish Succession in the first decade of the eighteenth century. The victories he won established Britain on the European stage as a force to be reckoned with; and the riches heaped upon him by a grateful nation allowed him to build a vast estate centred on the magnificent Blenheim Palace, named after his greatest victory. Having been made the 1st Duke of Marlborough, John Churchill founded a dynasty. However, like many grandee families over the generations, the Churchills experienced ups and downs, with later dukes exhibiting profligacy, instability and particularly poor management of their lives and resources, resulting in a huge sale of art treasures to keep the family solvent.[1] Winston's father, Randolph Churchill, was the younger son of the 7th Duke of Marlborough, and he was already pursuing a promising political career at the time of his elder son's birth. He had followed the recent example of several scions of the English aristocracy and married an American heiress, the charming and beautiful Jennie Jerome, whose wealthy father was a stockbroker and part owner of the *New York Times*.

By the accident of being eight weeks premature, Winston was born at Blenheim Palace.[2] His parents were on a shooting party there and Jennie was riding in a pony carriage over rough ground when she went into labour. Winston's earliest years were spent in London and Dublin. As was the custom at this time, he was brought up largely by his nanny, Mrs Everest, to whom he became devoted. His mother, whom he later described as a 'fairy princess', was remote but caring. He wrote, 'She shone for me like the Evening Star.'[3] His father, who was rising through the ranks of the Conservative Party and seemed to have a dazzling political career ahead of him, was even more remote and showed no signs of tenderness, despite his son's adoration and love. Churchill later commented that he had only three or four intimate conversations with his father during his whole life. When he was seven, the young Winston was sent to a brutal primary school near Ascot where floggings with birch were common. He was almost certainly bullied there as well. He hated the school, and after two years was taken away and sent to a much gentler establishment in Brighton.

Churchill wrote about his youth in *My Early Life*, published in 1930 when he was in his mid-fifties. It is a wonderfully entertaining account of how a backward pupil finds a niche in life. Churchill displayed little academic ability in the narrow sense in which it was defined in the late Victorian public school system: that is, in classics and mathematics. His description of taking his entrance examination to Harrow perfectly captures the hopelessness he felt in the face of exams and conventional learning. He was unable to answer a single question in the Latin paper and remembers:

> I wrote my name at the top of the page. I wrote down the number of the question '1'. After much reflection I put a bracket around it thus '(1)'. But thereafter I could not think of anything connected with it that was either relevant or

true. Incidentally there arrived from nowhere in particular
a blot and several smudges. I gazed for two whole hours
at this sad spectacle: and then merciful ushers collected
my piece of foolscap with all the others.[4]

From this slender indication of scholarship the headmaster of
Harrow nevertheless offered the young Churchill a place at the
exclusive school. It probably helped that his father was one of
the most famous Tory politicians in Britain at the time.

Churchill was no star pupil at Harrow, but while he was
hopeless at the conventional subjects, he had an extraordinary
ability to learn by heart, once winning a prize for reciting
twelve hundred lines of Macaulay's 'Lays of Ancient Rome'
word perfectly. And although he failed to absorb much Latin or
Greek, he did learn about the English language and how to
write a sentence. He showed a particular interest in history and
was skilled at writing essays in the subject (a talent that was not
much respected then, as perhaps now). Having been obsessed
with his toy soldiers, and disappointing his father because
he did not have the ability to go on to become a lawyer, it was
resolved that the young Winston should head for a career in the
army. Unfortunately, once again the problem of the entrance
examination loomed, this time to get into Sandhurst, where the
young officers-to-be of the British Empire were trained. This
time there was no favouritism to help a well-heeled son of one
of Britain's finest families into the officer class. Winston had to
get through the exams by himself, which included his old bug-
bear of mathematics. He failed the exams twice, then attended
a crammer school in west London. On his third attempt he just
scraped in – 95th out of 104 candidates. This was not high
enough to qualify for the infantry, but the cavalry had lower
standards and accepted him for a cadetship. At the age of eighteen, Winston Churchill was in the army at last.

Once at Sandhurst, Churchill's somewhat unpromising career took a completely new course. No longer handicapped by his lack of knowledge in Latin or mathematics, he began to enjoy courses in Tactics, Fortifications, Military Administration, Drill and Riding. He did well and soon stood out as good officer material and an excellent horseman. In December 1894 he succeeded in his final exams and passed out 20th in the list of 130. Then, with a little help from his mother and from the Marlborough family, he entered one of the most fashionable cavalry regiments, the 4th Hussars. They were smart, aristocratic and led by one of the leading officers in the British Army, a man who had close connections with the royal family. The only problem was that the salary of a young subaltern did not match the outgoings expected of an officer in this elite regiment, who had to provide his own uniform, run two horses and pay all of his mess bills. So any officer in the 4th Hussars, and indeed in most other cavalry regiments at the time, needed a private income. The bubble of Lord Randolph Churchill's career had burst when he resigned from the government in 1886 and, to his astonishment, was never asked back. Suffering from either syphilis or more probably some form of brain tumour, he experienced erratic mood swings and needed constant medical attention. By the time he died in 1895 he had used up all of his fortune. Winston's mother was left with barely enough to fund her own extravagant lifestyle, let alone those of her two sons. Consequently, money would be a problem for Churchill for some time to come, and the need to pay his own way partly determined his course of action over the following years.

Churchill threw himself into the life of his regiment. For an officer recruit this involved a round of activities at the Riding School, learning horsemanship; on the Barrack Square, learning cavalry manoeuvres; and in the mess, learning to be a true officer and a gentleman. At this point in the late Victorian era, the country had enjoyed many years of peace. Few officers below the rank of captain had seen any active service. It was Churchill's fear that he would serve dutifully for many years but not enjoy the thrill of combat. 'From very early youth I had brooded about soldiers and war,' he later wrote, 'and often I had imagined in dreams and day-dreams the sensations attendant upon being for the first time under fire. It seemed to my youthful mind that it must be a thrilling and immense experience to hear the whistle of bullets all round.'[5] So he now resolved to take full advantage of the perks of a young subaltern in a cavalry regiment, one of which was long holidays, and to put this matter right. In the winter of 1895, during his two-month break, instead of spending the time fox-hunting, as was usual for cavalry officers, Churchill and a fellow-officer travelled at their own expense to Cuba, where a war was raging between local rebels and the Spanish colonialists. With appropriate introductions from an old friend of Churchill's father, the two young officers were assigned to a mobile column of the Spanish Army marching into the jungle interior in search of rebels. They soon found them and a gunfight ensued in which the twenty-one-year-old Churchill came under fire for the first time. He found the whole experience exhilarating. But the mission was not a success. The Spanish forces deployed in conventional formation to assault the Cubans, who, adopting guerrilla tactics that would become much more familiar over the following hundred years, simply melted away into the jungle mists.

After a couple of weeks the column returned to base and Churchill and his friend sailed home. Today it seems incredible that a young army officer would pay his own passage halfway across the world to engage in combat, with all the risks of death or injury that entailed. But in the last decade of the Victorian era, before the futile horrors of the Great War, before the destruction of aerial bombing, and long before the nightmare of nuclear Armageddon came to haunt us, war was still seen as glamorous and romantic. Certainly Churchill saw it that way. And he was ambitious. The officers who had experience of warfare would probably be promoted more rapidly. Doubtless they would attract the awe and attention of fellow-officers. It also seems likely that Churchill was already looking ahead to a political career and wanted to notch up some worthwhile experience as a foundation for what would follow.

Soon after Churchill's return from Cuba, the 4th Hussars were sent to India. This was a regular posting for almost every unit in the British Army, and the Hussars were assigned to spend nine years in Bangalore in the south. For a young officer in an elite cavalry regiment, life on the India station in the heyday of the Raj could be very pleasant. Officers lived in spacious bungalows surrounded by neat gardens and were looked after by a butler and servants. There were a couple of hours of horse-riding drill from six o'clock each morning, then an hour or so in the stables, then nothing much through the heat of the day, until the officers started to play polo around 5 p.m. And each evening there were dinner and drinks in the mess. Churchill committed himself wholeheartedly to polo and soon developed into a fine player, despite sustaining a shoulder injury. But he rapidly realised that this leisurely officer's life was not enough for him. He needed something else.

Churchill was very aware that his earlier academic failings had forced him to miss out on a university education. So he decided he needed to catch up on his learning. In the many hours of his down time at Bangalore, he threw himself into a rigorous reading programme. His mother sent him crates of books which he devoured. He started with the eight volumes of Gibbon's *Decline and Fall of the Roman Empire*, which someone

told him had been a favourite of his father, and the twelve volumes of Macaulay's *History of England*: 'fifty pages of Macaulay
and twenty-five of Gibbon every day'. He then progressed to
other classics of history and philosophy, from Socrates to Mal-
thus and from Aristotle to Darwin and Adam Smith. He even
asked his mother to send dozens of volumes of the *Annual
Register*, a compendium of parliamentary debates and an official
record of British public life. He read for four or five hours every
day, for five or six months of the year. A mind that had not been
accustomed to learning was suddenly soaking up ideas like a
sponge. He loved the way the English language was used in
these classics and was absorbed by the stories they related and
the ideas they contained. And he stored away everything he discovered. His scholarly reading must have made him a very
unusual figure among the other young cavalry officers of his
regiment. But his enthusiasm for polo kept him in with his
fellow-officers as a popular and sporting colleague.

Churchill longed for one of India's regular frontier wars in
which he could seek further experience and possible fame. But
in sleepy Bangalore all he had were his books, his polo and the
daily round of regimental life. Then, in the spring of 1897, a dispute arose in the Swat Valley in Malakand on the North-West
Frontier (now still an unruly quarter of northern Pakistan).
Churchill was on leave in England but immediately raced back
to India to try to be assigned to the field force that was setting
out to teach the Muslim Pathan rebels a lesson. The commander
cabled him: 'No vacancies; come as a correspondent; will try to
fit you in.' Churchill rushed first to Bangalore, to get permission
to join the field force, and then travelled for five days by train
to the North-West Frontier. Meanwhile, back in London, his
mother lobbied various editors and finally persuaded the *Daily
Telegraph* to accept dispatches from her son at five pounds a
column.

The Malakand Field Force was a unit typical of the British Raj. It consisted of regular British Army units on their tour of duty in the subcontinent, and units from the Indian Army, with British officers commanding native warrior-soldiers – Sikhs, Punjabis and others. Travelling with the force was a set of 'political officers' whose job it was to negotiate with the locals and enforce imperial rule. The field force's mission was to seek out the Pathan warriors and draw them into battle. On 16 September a small group was detached to go up the Mamund Valley. Churchill was advised that he might see some action here, so he joined the 35th Sikhs, who slowly marched their way up the valley, surrounded by mountains rising steeply to four or five thousand feet. At the top of the valley they reached a village. The troops were about to destroy the villagers' crops as a form of collective punishment when Churchill looked around and realised there were only four or five officers and about eighty Sikhs. The rest of the column was way behind them down the valley. At that moment, firing erupted and Churchill could see the glint of the swords of the Pathan tribesmen reflecting in the hot sun along the steep valley side. They had walked into a trap.

Churchill picked up a rifle and began to return fire. A British officer ordered the small force to withdraw down the valley. He was shot and killed only a few yards from Churchill. The Sikhs pulled back in some confusion and nearly broke one of the first rules of a frontier war: never leave the wounded behind at the mercy of an enemy who would probably hack them to pieces. But Churchill and a few of the Sikhs carried their wounded comrades down the valley, under constant harassment from groups of Pathan warriors. At one point a tribesman charged at Churchill, brandishing his sword. Churchill took out his revolver and fired. He missed, but the warrior withdrew hastily and hid behind a rock. Eventually, Churchill and his paltry force reached the rest of the company further down the valley, but the tribesmen were still in hot pursuit. Then came the reassuring sound of regular firing and the smart order 'Volley firing. Ready. Present' echoed across the valley. Another volley of rifle fire crashed out. A regular British Army unit, the East Kents, known as the Buffs, had arrived on the scene to save the day.

After further intense fighting, the numerically superior British and Indian troops finally took control of the valley. Over the next two weeks Churchill witnessed the systematic destruction of the Pathans' villages, the filling in of their wells, the burning of their crops and the smashing of their reservoirs in punishment. Such was the revenge of the British Empire. But for Churchill this combined Anglo-Indian expedition confirmed his belief in the Empire and his conviction that Britain had a mission to rule India. It was a belief he never gave up.

Churchill's dispatches in the *Daily Telegraph* were well received for their graphic and dramatic accounts. Encouraged by this, he wrote a book of the campaign which he sent back to London and his mother arranged for its publication. *The Malakand Field Force* was a great success, well reviewed and widely read. Even the Prince of Wales wrote to congratulate Churchill: 'Everyone is reading it, and I only hear it spoken of with praise.'[6] Churchill reflected that for a few months' hard work writing the book he had earned the equivalent of two years' pay as a cavalry officer. He was delighted with the praise and took on board the pecuniary lesson.

A year later another imperial sortie attracted Churchill's attention. Lord Salisbury's Conservative government had decided to send an army to the Sudan to teach the Khalifa and his Muslim Dervish army a lesson for the assault by his predecessor on Khartoum a decade earlier, which had cost the life of the British commander there, General Gordon. Sir Herbert

Kitchener assembled an expedition to march down the Nile into the Sudan and on to the Dervish capital. Churchill again tried to pull strings with influential people to get himself assigned to an expedition that offered even more dramatic imperial adventure than the North-West Frontier. However, despite much support for his placement, Kitchener refused to have Churchill in his expedition. There was clearly some hostility felt towards the young cavalry officer who always seemed to up sticks and leave his own regiment to be at the centre of the action. And many people did not like the idea of such a junior officer going into print and criticising his superiors. Even a telegram from the Prime Minister's office did not change Kitchener's mind. But Churchill was nothing if not persistent, and at the last minute the War Office assigned him to the 21st Lancers, who were to accompany the expedition south. Churchill embarked immediately for Cairo, where he joined the Lancers just as they were leaving. The border with the Sudan lay some fourteen hundred miles to the south. This time Churchill was contracted to supply letters to the *Morning Post* at fifteen pounds a column. His value was rising.

As we saw in the Introduction, Churchill marched with Kitchener's army down the left bank of the Nile, then acquitted himself bravely in the Battle of Omdurman. Just as on the North-West Frontier, he looked death in the face from frenzied tribesmen opposed to British rule and once again emerged without a scratch. Once this short engagement was over, 'the most dangerous 2 minutes I shall live to see', as he wrote to a fellow-officer, Churchill played no further role in the campaign.[7] Three hundred British troops died at Omdurman. But about ten thousand Dervishes were killed and another fifteen thousand were wounded. The wounded were left to die in the hot desert, and were offered no medical aid. Some were even murdered where they lay. Later, when British and Egyptian soldiers entered Khartoum, they desecrated the tomb of the Mahdi who had destroyed Gordon's army ten years before. His corpse was exhumed, decapitated and eventually taken to Cairo. Churchill was deeply shocked by this.

He returned home within days of the victory at Omdurman. The adventure was over. He briefly rejoined his regiment in Bangalore and helped them to win the Indian Polo Championship, despite a worsening of his shoulder injury. Then, having achieved his objectives of seeing action and commanding men under fire, he resigned his commission. The pace of regular army life was too slow for him and he wanted to move on. He returned to London to devote himself to writing and politics. Free to write without the limitations of being an army officer, he quickly finished another military book which told the story of the Sudan campaign, *The River War*. Published in two volumes and at 950 pages, this was as successful as his earlier work and helped to enhance his reputation. In *The River War* Churchill was outspoken in his criticism of Kitchener for failing to prevent the brutality of the army once the victory had been won. He was already beginning to work out his own philosophy of war, which involved being defiant in the face of defeat, resolute in the pursuit of battle, but magnanimous in victory. Kitchener's army had shown only barbarity after its triumph, and Churchill argued that this was not right. Unsurprisingly, the book won him few friends in military circles.

Back in London, Churchill decided it was time to launch his political career, and sounded out the Conservative Party.[8] The mixed reputation of his father preceded him and he was made a candidate for the tough working-class constituency of Oldham. The Lancashire cotton town faced a by-election in the summer of 1899. Churchill fought his first election campaign and lost. He felt disconsolate at his defeat. But, as ever in his remarkable life, unexpected events came to the rescue.

In October 1899, just as *The River War* was about to be published, war broke out with the Boers in South Africa. The quarrel between the ever-expanding British Empire and the Boer republics went back a long way. The Dutch settlers who had lived in southern Africa since the seventeenth century had been slowly migrating north from their old colony at the Cape. They were fiercely independent, strongly Calvinistic and had fought a succession of minor wars with their neighbours, the Zulus, as well as the British. What transformed a series of relatively petty arguments into something far more significant was the discovery of huge reserves of gold and diamonds in the Transvaal in the 1890s. Now that there were abundant natural resources to be exploited and fortunes to be made, the warlike Boers attacked British territory. For the first time in nearly half a century, Britain found itself at war with other white men. It was soon clear that this could escalate into a major imperial conflict.

Churchill once again felt the magnetic pull of war, and he negotiated a deal with the *Morning Post* to report from South Africa as a war correspondent. This time he commanded the princely fee of £250 per month plus all expenses, making him one of the highest-paid war reporters in South Africa. He sailed on the first available boat for Cape Town, travelling with the British commander-in-chief, General Redvers Buller, and his staff. Churchill's dispatches from South Africa would soon turn him into an international celebrity.

As with so many wars fought by Britain, the early stages of the Boer War went badly. The Boers showed themselves as skilled fighters, good horsemen and ingenious tacticians, and they had bought the latest weaponry, including magazine-loading rifles and modern artillery pieces. They laid siege to the British garrisons at Mafeking, Ladysmith and Kimberley, and in almost every head-on engagement they proved superior to the British forces. Churchill had to report one setback after another. In an attempt to be first with his story, he negotiated a trip on an armoured train from Durban to Estcourt. Then, on 15 November, the train ventured forth from Estcourt with two companies of infantrymen under the command of Captain Haldane, who asked Churchill to go along with them. The train moved cautiously into territory that had recently been raided by the Boers. On its way back it came under fire from a Boer raiding force and then ran into an ambush. The Boers had placed a heavy stone on the track and three coaches were derailed.

Churchill had only the status of a reporter, but so recently out of uniform and now finding himself and his companions under fire, he soon reverted to military mode. With Haldane's agreement, he spent about an hour trying to get the steam engine to push the derailed coaches out of the way. Throughout this time they were under continuous rifle and artillery fire. Four men were killed and about thirty wounded. Haldane wrote in his official report that Churchill 'with indomitable perseverance continued his difficult task'.[9] Eventually, Churchill managed to get the engine past the derailed coaches and it carried off the wounded men to the nearest town and safety. Churchill himself went just a few hundred yards and then left the locomotive to walk back to the scene of the ambush, where Haldane and the remaining soldiers were still exchanging fire with the Boers. He had not gone far when two Boer riflemen surrounded him and opened fire. Churchill ran back down the railway cutting with the riflemen shooting after him. 'Their bullets,' he later wrote, 'sucking to right and left, seemed to miss only by inches.'[10] He emerged from the cutting and headed for the cover of a river gorge, but was chased by a Boer horseman. With the Boer just forty yards behind, Churchill reached for his Mauser pistol. Later he said that, with his blood up, he would have killed the horseman. But his revolver was not there. He had taken it off earlier. The Boer aimed his rifle at Churchill who now had no alternative. He put his hands in the air and surrendered. Churchill was a prisoner.

All the men who had surrendered were rounded up like cattle and taken away. This action was typical of the many humiliations which the Boers inflicted on British forces at this stage of the war. The technically superior firepower of the armoured train had counted for nothing against a cleverly planned ambush by able fighters who had chosen their ground well and could soon overpower the British troops.

Churchill and the officers were taken to a school that had been requisitioned as a prisoner-of-war camp in Pretoria. When captured, Churchill had been unarmed and he had all his journalistic credentials on him. But the Boers realised they had a big catch in Winston Churchill and were unwilling to release him, despite his protests that he was an unarmed civilian. Churchill wrote formally to the Boer commander, demanding to be released. He claimed that he had at no time fired on Boer forces and had only been trying to evacuate the wounded. His appeals were ignored.

Churchill passed his twenty-fifth birthday in the POW camp. Thirty years later he wrote, 'I certainly hated every minute of my captivity more than I have ever hated any other period of my whole life.'[11] There is a photo of him at the camp in which he looks dejected and thoroughly peeved. When it seemed that his appeals to the Boers would fail, he began to plan an escape. He, Haldane and another prisoner who spoke some Dutch intended to climb over the hastily built prison fence and head east out of Boer territory into neutral Mozambique, travelling by night and resting up during the day. After several delays, on the night of 12 December Churchill made a dash for it while the guard was not looking. He clambered over the fence and into a neighbouring garden, where he waited in the shrubs for the others. But no one came. The guard was too watchful and the others could not make their escape. Churchill was on his own. He had some money in his pocket and some chocolate, but no map.

Unknown to Churchill, when his absence was discovered a huge hue and cry went up. Search parties were sent out to look for him. Posters were put up offering a reward of twenty-five pounds, 'dead or alive'. But he managed to hide on a train heading east through the night and the next morning he sheltered near a tiny station. He was safe for the moment, but knew that without food or help he would never be able to find his way the three hundred miles to Mozambique. There were guards on every bridge and at every station. He could not decide what was best. 'I stopped and sat down,' he later wrote, 'completely baffled, destitute of any idea what to do and where to turn.'[12] Finally, he took a huge risk and went up to the door of a house in a nearby kraal, or settlement, to ask for help. With extraordinary luck, he had picked the house of a British engineer who managed the local coal mine. The engineer knew that Churchill was a wanted man, but still he decided to help. He fed him and hid him in the mine for several days while he made a plan. With the support of three others (one of whom was from Oldham and so knew of Churchill's recent by-election campaign), the engineer then smuggled Churchill into the truck of a coal train heading east. Churchill eventually reached Lourenco Marques (now Maputo), where, covered in coal dust, he walked into the British Consulate and freedom.[13]

Churchill's escape brought him instant fame. It provided a brief moment of relief and celebration at a point when the war was going badly for the British. When Churchill arrived back in Durban on 23 December he was met by cheering crowds, who took over an hour to disperse. Three days later he met General

Buller, the commander of the British forces in Natal, who congratulated him wholeheartedly. Churchill had become a hero. And now he asked to enlist again in the fighting forces.

This posed a quandary for Buller, because the War Office had recently made it clear that serving officers could no longer write for the press and no journalists could fight in the regular army. But Buller was so keen to enlist Churchill with his fighting spirit that an exception was made and he was offered an unpaid commission in a local regiment, the South African Light Horse. He spent the rest of his time in South Africa with the unusual dual role of being both a fighting cavalry lieutenant and a war correspondent.

In the early months of 1900, Churchill witnessed some of the worst moments of the Boer War, including the aftermath of the fighting at the Tugela River and the battle at Spion Kop. For many days at a time he lived under constant shellfire and regular rifle fire. At one point, the feather in his hat was cut through by a bullet. At another, eight men around him were wounded by a shrapnel burst while Churchill was, yet again, unscathed. In April, he found himself alone, facing a group of Boer commandos, cut off and without a horse. At the last minute, a British scout rode up and Churchill leaped on his horse. They rode off together, but the horse was shot and died of its wounds. 'I don't think I have ever been so near destruction,' he wrote to his mother.[14]

These months of fighting and writing reports for the *Morning Post* helped to shape Churchill's view of war. He was fascinated by the commando tactics of the Boers, who attacked in small numbers, struck hard and then melted away into the countryside. This left Churchill with a lifelong respect for the use of small, well-trained forces that could hit the enemy with an impact out of all proportion to their numbers. This appealed to his romantic view of war and how it should be fought. He also later became firm friends with some of the Boer commanders, men like Louis Botha and Jan Christiaan Smuts.

On the other hand, although he continued to find the thrill of battle exhilarating, he was terribly moved by the death and mayhem he witnessed in South Africa. After the fighting on the Tugela River, he came across the dead bodies of two Boers: a man in his sixties who had been wounded in the leg and had bled to death; and alongside him a young boy of about seventeen, shot through the heart. A few hundred yards away were the corpses of two British soldiers, their heads smashed 'like egg shells'. Churchill wrote in the *Morning Post*: 'I have often seen dead men killed in war – thousands at Omdurman, scores elsewhere, black and white, but the Boer dead aroused the most painful emotions . . . Ah, horrible war, amazing medley of the glorious and squalid, the pitiful and the sublime.'[15]

Churchill also refined his views about the senior officers in the British Army. He felt they were not facing up to reality and were not using their troops effectively. He was amazed at how officers still ordered men into frontal attacks against troops who were able to employ the devastating firepower of their powerful modern rifles. There was no glory, only sacrifice, in this. He wrote in the *Morning Post*: 'We must face the facts. The individual Boer, mounted in suitable country, is worth from three to five regular soldiers. The power of modern rifles is so tremendous that frontal attacks must often be repulsed.'[16] He also felt that senior officers rarely showed enough aggressive spirit and too easily became depressed and almost fatalistic about the fact that the enemy would outperform them. This sense that British commanders needed to be more offensive-minded would return to worry Churchill again, forty years later.

In January 1900, Lord Roberts took command of the British Army in South Africa and Kitchener was sent from Khartoum as his chief of staff. With reinforcements of men and supplies arriving from Britain, the tide of war slowly turned against the Boers. Churchill wrote that the British now needed to show mercy to the Boers so as not to provoke a further phase of bitter warfare. His views were unpopular, going against the grain of jingoistic fervour that had been stirred up back home. Nevertheless, this was an important aspect of his concept of the morality of war – to be magnanimous in victory and to show goodwill in peace.

In May, the siege of Mafeking was relieved, prompting huge celebrations in Britain, and the creation of another new hero, Major Robert Baden-Powell, who had led the town's garrison through the dreadful hardships of the siege. In June, Churchill was present at the recapture of Pretoria, leading the troops that freed the remaining prisoners-of-war. Amid cheers, he tore down the Boer flag, replaced it with the Union Jack, and was reunited with many of those who had been captured with him six months before.

With the recapture of Pretoria and the relief of Mafeking, it seemed that the war had been won, and many people assumed it would soon be over. In fact, it dragged on for a further two years, with the Boers mounting a highly effective guerrilla war against the British right across southern Africa. Exasperated, the British, by now under Kitchener's command, did everything they could to destroy ground support for their guerrilla enemy. Farmsteads were destroyed, crops were burned and Boer citizens were rounded up and interned in what the British called 'concentration camps'. Tens of thousands of Boer women and children died in the overcrowded and unhealthy camps, leaving a lasting legacy of hatred towards British rule among the Afrikaner community.

Churchill returned to Britain in the summer of 1900 and immediately resumed his political career. The Conservative government of Lord Salisbury wanted to exploit the patriotism generated by the war and called a general election in the autumn. It became known as the 'Khaki Election'. Churchill once again stood in Oldham, where he was welcomed as a returning war hero by a crowd of ten thousand people. During the campaign, he was cheered wildly in speech after speech. In those days, general elections were not held on a single day, with the result announced the following morning. Instead, the election process could last anything up to five weeks. However, Oldham was one of the first constituencies to declare its result, and it returned two members. This time Churchill was elected as the second candidate by a narrow margin. This early victory in a working-class town gave a huge boost to the Conservative campaign, and Churchill was in great demand during the remaining weeks of voting. He addressed political rallies up and down the country, every night for four weeks, sharing the platform with many of the leading Conservatives of the day.

The triumphant Churchill, basking in the glory of his personal political success, then went on a whistle-stop lecture tour, using a magic lantern to tell the story of his experiences in the Boer War. Today, a young celebrity like the twenty-six-year-old Winston Churchill would probably go on a reality TV show. Then, public speaking was the way to become famous and to earn money. Churchill visited most of the big cities of Britain, captivating crowds of many thousands at a time, and often earning a hundred pounds or more for an evening lecture. The House of Commons, in which the Conservatives and Unionists enjoyed an increased majority, was due to meet in December, but the newly elected MP for Oldham chose to miss the opening of his first Parliament and instead continued his lecture tour in the United States and Canada. Here, even more money was on offer. After three lucrative months, Churchill had amassed about ten thousand pounds (roughly a million in today's money). He invested his earnings and had plenty to live on for years to come. The financial worries that had plagued him since he had joined the army were over and he was free to dedicate himself to his new political career. His military exploits had led to fame *and* fortune.

The young MP gradually began to build a new reputation for himself at Westminster. Fast-forwarding through his political career, which is not the subject of this study, Churchill never felt entirely comfortable within the Conservative Party. In particular, as a convinced free-trader, he disagreed with the Tory policy of tariff protection. In October 1903, he drafted a letter to a friend. It was never sent, but in it Churchill stated, 'I am an English Liberal. I hate the Tory Party, their men, their words and their methods.'[17] In May 1904, he took the highly unusual step of 'crossing the floor' of the House of Commons; that is, he left the Conservative benches and joined the Liberals. Many Tories never forgave him for this act of betrayal. And as the Conservative government was becoming unpopular, many others saw it as a purely opportunistic act, an attempt to seek office within a future Liberal government.

Sure enough, two years later, Sir Henry Campbell-Bannerman's Liberals won the general election. Churchill was elected as the Liberal MP for Manchester North West. With his fame and celebrity status, he was an obvious candidate for office and he duly became Under-Secretary of State at the Colonial Office. For two years he threw himself into his new role as a junior minister and oversaw the creation of self-governing states in the Transvaal and the Orange Free State, bringing the Boers into the British Empire and resolving the disputes in southern Africa. At one point he wrote six lengthy notes for the Colonial Office, outlining his plans for various other parts of the world. This prompted Sir Frances Hopwood, the senior

civil servant at the Colonial Office, to write, 'He is most tire-
some to deal with & will I fear give trouble – as his father did –
in any position to which he may be called. The restless energy,
uncontrollable desire for notoriety & the lack of moral perception make him an anxiety indeed.'[18] So Churchill was already
displaying the energy and drive that would characterise his
wartime leadership, but was still dismissed as an awkward
troublemaker by many in the establishment.

In the year 1908, two life-changing events took place. In
March, Churchill met Clementine Hozier at a dinner party. She
was ten years younger than him, and radiant. Churchill particularly liked her striking, mysterious eyes. He asked her if she
had read his recent biography of his father. She had not. Despite
this, Churchill became infatuated with Clementine and they
wrote and met regularly over the next few months. This was
not Churchill's first infatuation, there had been a small number
of society women who had previously attracted his attention.
But he was the first to admit that he was not a great romancer,
and found it difficult talking to young ladies. Clementine,
though, was different: she was serious-minded as well as beautiful; and, rarely for a girl of her class at the time, she had good
academic qualifications.[19] She liked his style, his wit and no
doubt his ambition, and when he proposed to her at Blenheim
Palace in August, she accepted. Churchill was delighted and the
two were married the following month.

Churchill loved 'Clemmie' intensely for the rest of his life and
he never strayed. She provided the support, homeliness and
large doses of good sense that he desperately needed. The
domestic life they began to build together was something new
for Churchill, who had endured a lonely childhood and since
Sandhurst had found most of his camaraderie in the male
worlds of the army and the House of Commons. Clemmie was
the ideal political wife. Despite long separations when Churchill

was away on business, they wrote lovingly to each other almost daily, and Churchill found Clemmie's support, always imbued with a great deal of common sense, a vital prop to both his emotional and his political life.[20]

While Churchill was courting Clemmie, Herbert Asquith replaced Campbell-Bannerman as Prime Minister. In the reshuffle that followed, Churchill was appointed President of the Board of Trade. So, aged just thirty-three, he became a member of the Cabinet. His star was rising quickly, along with that of another passionate and visionary politician, Chancellor of the Exchequer David Lloyd George. Asquith's government would become one of the greatest social-reforming administrations of the twentieth century, laying the foundations for the welfare state. Churchill was soon hard at work drafting legislation to create a minimum wage, to establish workers' rights to breaks for meals and refreshment (the much-loved British tea break became law in one of his reforms), and to create more than two hundred labour exchanges across the country to help the unemployed find work. Always keen to get into print, Churchill's next book was a compendium of his speeches on reform entitled *The People's Rights*.

In February 1910, Churchill was promoted again, this time to Home Secretary. He was now even more centrally placed to carry forward the Liberal agenda of improving conditions for working people. He also threw himself into prison reform and reduced the high numbers of young offenders in prison. But the tensions of Edwardian Britain were never far below the surface. There was the long-running and still unresolved issue of Home Rule for Ireland. Churchill was lukewarm in his support for Home Rule, even though it was official Liberal policy. British society was similarly divided over the issue of whether women should get the vote (there was still not universal suffrage for men, either). The suffragette movement eventually split, with

one group resorting to violent protest in a bid to make its voice
heard. The chant 'Votes for Women' echoed around Westminster. Churchill was not in favour. Clementine was, but was
opposed to the violent tactics of the militants. As Home
Secretary, Churchill approved the forced feeding of suffragettes
who had gone on hunger strike in prison.

Along with this, Britain was hit by a series of increasingly
damaging labour disputes, as working men and women began
to exercise their political muscle. On Churchill's watch there
was a series of strikes in the South Wales coal mines which led
to local rioting and disturbances. Shops in the town of
Tonypandy were looted and a local colliery attacked. In principle, Churchill was opposed to the use of the army to resolve
domestic disputes, declaring in the House of Commons: 'It
must be an object of public policy, to avoid collisions between
troops and people engaged in industrial disputes.' However, he
sent a squadron of cavalry to the Valleys and placed them on
stand-by. The soldiers were never used, but Churchill was still
widely condemned from all sides. The Conservatives accused
him of being too soft on the rioters. The Labour Party, although
still only small as a party at Westminster, was outraged that he
had sent soldiers into an industrial problem. 'Remember
Tonypandy!' was a cry heard against Churchill from the Left for
many years to come.

These were all serious issues, but Asquith's administration
faced its greatest crisis yet when the House of Lords rejected
Lloyd George's reforming budget, known as the 'People's
Budget'. In this, he proposed unemployment benefits and, most
radically, Britain's first state pension. In order to pay for these,
Lloyd George planned to increase taxation of the rich, including a new super-tax of six pence in the pound for those earning
over five thousand pounds per year, along with rises in death
duties and property taxes. The Conservatives were deeply

opposed to this attempt to redistribute wealth, and used their substantial majority in the House of Lords to reject the budget. Asquith was outraged that the Lords could reject a money bill proposed by a democratically elected House of Commons, and a major constitutional crisis unfolded. Two general elections followed in 1910, the House of Lords finally gave in, and the budget was passed. But Asquith had not finished: he wanted to permanently restrict the power of the House of Lords and he persuaded King George V to agree in principle to the creation of 250 new Liberal peers if the Lords did not accept another piece of legislation, the Parliament Bill. In August, to avoid being swamped by the new peers, the Lords finally passed the bill by a tiny majority. The constitutional crisis was over. The People had won. Churchill, firmly committed to the side of the People, was viewed by the Conservatives as no less than a class traitor. They were now more hostile towards him than ever.

A totally different crisis also erupted in the summer of 1911 when the Germans sent a gunboat to Agadir in Morocco during a revolt against the Sultan. The French regarded Morocco as part of their sphere of influence in North Africa and were appalled at what they saw as aggressive German action. The British interpreted this piece of 'gunboat diplomacy' as an attempt to turn Agadir into a German port, a clear threat to the British naval base at Gibraltar, as well as a sign of Germany's ambition to rival the Royal Navy. The crisis was soon defused, but the Prime Minister decided that the Admiralty needed someone more assertive in charge. In October 1911, Asquith invited Churchill to visit him in Scotland. After a round of golf, the Prime Minister asked his Home Secretary quite abruptly if he would like to become First Lord of the Admiralty. Churchill later wrote that, after the crisis that summer, 'All my mind was full of the dangers of war. I accepted with alacrity.'[21] It was a few weeks before his thirty-seventh birthday. His new role would shape Churchill's career and his military thinking for years to come.

The Royal Navy, over which Churchill took civilian and political command in October 1911, was a mighty force that still ruled the waves. However, like many aspects of pre-Great War Britain, its supremacy was severely challenged by a series of weaknesses and fissures, some evident to contemporary observers, others less visible. A naval race had begun when Germany, traditionally a friend of Britain (the Kaiser was the nephew of King Edward VII), started to expand its fleet. British policy for almost a century was to possess a small, professional army but a vast navy that could defend British imperial and trading links around the world. The Royal Navy was intended to be as large as the next two most powerful navies combined. So when Germany started to build up her fleet this was taken as a major affront, an attempt to diminish Britain's authority in world affairs. Along with this rivalry came the development of an entirely new generation of fast-moving, turbine-powered, heavy steel-clad battleships called Dreadnoughts. This new class of super battleship left most of its predecessors obsolete. So, as the Germans began to build Dreadnoughts, the British government needed to do the same, at enormous expense, to maintain its supremacy. Earlier, in Cabinet, Churchill himself had opposed the cost of this in order to keep funds for his social reforms.

Once again, Churchill threw himself into a new challenge. Although he had served in the army and had written extensively about military campaigns, he knew relatively little about the navy. But he was eager to learn, and through long discussions with his chief naval advisers, the Sea Lords, along with extensive visits to naval bases and meetings with junior personnel, Churchill began to pick up knowledge of the intricacies of naval gunnery, the relative merits of different vessels, and the key elements in the complex organisation of the Royal Navy. The Admiralty at this time put at the disposal of the First Lord a 320-foot yacht, the *Enchantress*, with a crew of 196. Over the next two and half years, Churchill spent more than two hundred nights on board the yacht, witnessing reviews and generally trying to understand naval matters. In other words, he spent nearly one night of every four at sea.

One of Churchill's first tasks was to push through a major naval reform by creating a Naval War Staff. Indeed, he prepared a paper for the Cabinet on this subject within four days of his appointment. The Naval War Staff would be a central team of officers to look across the board at the threats facing Britain's sea power, and then establish ways of dealing with them. Its creation was based on reforms that had taken place five years earlier in the army at the War Office. The growing threat from German naval expansion made this reform a necessity, but it involved a level of strategic thinking that the Royal Navy was not used to. For the navy, promotion had always come after service at sea, and experience in the various ships of the line was the most favoured knowledge in Admiralty thinking. The First Sea Lord, Sir Arthur Wilson, the most senior naval officer, was deeply opposed to this change in naval tradition. He feared the creation of a new cadre of staff officers who might get to the top without devoting their lives to service at sea. Churchill soon removed Wilson and the Naval War Staff was created. Then, typically, Churchill himself wanted to be at the centre of this strategic review of naval threats and opportunities.

Fresh from his days as a social reformer, Churchill also wanted to improve the lot of the general sailor. He improved pay as well as facilities below decks and at shore establishments. And his 'Mates' Scheme' enabled ratings to be promoted to officer rank. Churchill wrote and communicated directly with officers below the rank it was regarded as appropriate for a First Lord to deal directly with. The Sea Lords deeply disapproved, but Churchill used it as a way to find out what junior officers were thinking and what they were concerned with, all of which he regarded as part of his remit.

He soon forgot his earlier opposition to the increased spending requirements of the navy. Alarmed by the growth of Germany's fleet, he speeded up the building programme of the mighty Dreadnoughts. In early 1912, he received Germany's new plan for naval construction. It projected huge growth, from the biggest new battleships to much smaller vessels, and a vast increase in naval personnel. Churchill calculated that at present growth rates the German Navy would one day deploy twenty-five battleships in the North Sea, whereas the Royal Navy would be able to put only twenty-two to sea. This would result in a substantial shift in the balance of power. It could not be allowed to happen. Churchill committed his energies to persuading his colleagues in Cabinet and throughout the nation of the urgent need to speed up the building of new Dreadnoughts. As he later put it: 'The Conservatives wanted six, the Liberals wanted four; we compromised on eight.'[22]

Churchill found many of his senior naval advisers rather stuffy, plodding and distinctly poor in analysis. The qualities that made a great naval captain were combative arrogance, the confidence to take risks and to be highly individualistic in the assessment of a situation and the leadership of men. But these were not the characteristics that made for good managers. Churchill later wrote of the senior admirals: 'They are so cocksure, insouciant and apathetic.'[23] Moreover, few senior naval men had the capacity to sit down and argue a case with an experienced debater like Churchill. They could not pull an argument to pieces and put it back together again. They often wilted under a concerted argumentative assault from Churchill, who thought them all the poorer for this. To be fair, of course, these were not the qualities that were judged to be admirable in naval circles.

One man who did delight Churchill was the retired admiral Sir John Fisher, who had been a controversial First Sea Lord from 1904 to 1910. He was ebullient, energetic and constantly looking for improvements, to find faster ships, deadlier weapons and better ways of doing things. It was Fisher who had overseen the introduction of the Dreadnoughts. He was a genuine eccentric even in the unusual circle of senior admirals, and Churchill took to him immediately. In many ways, the two men were alike. They both had an instinct to spot what was new and different. And each identified in the other a keen supporter of his own passionate beliefs. Fisher was soon encouraging Churchill in his reforms, and the two men enjoyed a close friendship. Inevitably, they argued, but this only strengthened their relationship. Churchill was keen for people to stand up to him and it was this that he missed in most of the senior admirals around him.

Another senior figure whom Churchill admired was Prince Louis of Battenberg, a cousin of the King. The royal family was close to the senior service. The recently crowned King George V had served in the navy and both of his sons, the future Edward VIII and George VI, were educated at the Royal Naval College. Churchill had a lively correspondence with the King about naval matters, especially the naming of new ships. For instance, the King vetoed the name *Oliver Cromwell* for a battleship, but Churchill got approval for his suggestion of *Iron Duke*. Churchill appointed Prince Louis as First Sea Lord in December 1912. He served Churchill and the navy well.

Churchill initiated several reforms in the years before the First World War. First, he gave great encouragement to the development of the Royal Naval Air Service. The Wright brothers had ushered in the era of powered flight only a few years earlier, in

1903. And the first flight did not take place in Britain until 1908, only three years before Churchill went to the Admiralty. But he was fascinated by this new activity, and although it is unrealistic to claim he spotted its full military potential at this stage, he certainly encouraged the navy to take up flying and to build a series of air stations. At this point, air power was seen simply as a form of reconnaissance, a potential extension of the 'eyes' of the navy to spot its enemy at sea. So keen did Churchill become that he asked some young naval pilots at their base at Eastchurch if they would teach him to fly. The pioneer aviators must have been astonished at the First Lord's enthusiasm, because flying was so dangerous that one flight in every five thousand resulted in a death. Nevertheless, Churchill persevered with his flying lessons throughout 1913, despite fierce opposition from his friends, his family and from his wife Clemmie. In December, after one of his instructors was killed in an accident, F.E. Smith, the politician and close friend of Churchill, wrote to him: 'Why do you do such a thing as fly repeatedly? Surely it is unfair to your family, your career and your friends.'[24] But Churchill's obstinate streak came through and he persisted with his flying lessons for another six months. Then, one of the planes he had flown crashed into the sea, killing the pilot who had been teaching Churchill only days earlier. Clemmie, who was five months pregnant with their third child, pleaded again with him to give up the deadly sport. This time he agreed and, despite nearly gaining his pilot's licence, reluctantly abandoned his flying lessons.

However, Churchill's commitment to naval flying only intensified. He made plans to draw civilian fliers into the navy. He named the type of aircraft that could land on water a 'seaplane' and ordered a hundred of them. He planned and budgeted for the building of five new air stations and for the provision of new flying facilities. And he was impressed by a flight he took in an airship over Chatham dockyard. Churchill's recurring fascination with the new attracted him to flying and drove him to push through these measures against an inherently conservative Admiralty establishment that just didn't get it.

The other major area of reform ushered in by Churchill at the Admiralty was the launch of another new class of battleship, the Queen Elizabeth. At twenty-five knots they were faster than anything that had gone before. They were armed with giant fifteen-inch guns, the largest in the navy. And, in a revolutionary step, they were powered by oil rather than coal. To support them, Churchill began a process of converting the whole navy to oil power, a massive transition that would take decades to complete and involved building oil storage depots around the world. Coal had been at the heart of Britain's industrial revolution and its supply was guaranteed for years to come, but oil was more efficient and lighter. As part of the transition, Churchill recognised the need to ensure that oil supplies would be secure well into the future. Consequently, in June 1914, in one of his most far-sighted acts, he persuaded the government and the House of Commons to take a 51 per cent stake in the Anglo Persian Oil Company (which later became British Petroleum). This marked several major shifts in long-term government thinking. First, it guaranteed the Royal Navy a plentiful supply of oil for the foreseeable future, enabling the transfer from coal to oil to proceed smoothly. Second, it marked a break from governmental laissez-faire policy towards business by creating a partly nationalised company dedicated to the provision of an essential raw material. For decades, the interests and investments of BP would be closely aligned to the interests of the British government, in marked contrast to the American oil giants that became huge private concerns. Third, the stake in Anglo Persian focused British attention on a region that had not been paramount in imperial thinking before: Persia (now Iran) and the Middle East. This would have major consequences over

the next half century. It was one of the most radical steps taken
by the pre-war government, and it is a sign of Churchill's
achievement that, despite being so revolutionary, it was carried
in the Commons overwhelmingly by 254 votes to 18.

Eleven days after the Commons voted to take the stake in
Anglo Persian, a shot rang out in Sarajevo. This started the
sequence of events that would lead Europe inexorably to war.
But at the time few realised the importance of the assassination
of Archduke Franz Ferdinand of Austria by a Serbian gunman,
so great were the issues facing the British government. The
Home Rule Act threatened to provoke civil war in Ireland.
The Unionists in the North rallied behind their leader Sir
Edward Carson. The Nationalists mobilised too, and groups
of armed men marched openly in Belfast and Dublin. The
suffragettes brought more violence to the streets of London.
Labour disputes were causing real anxiety throughout the
country. And the Cabinet was still arguing over the heavy
expenditure demanded by the navy. When the Naval Estimates
were finally agreed for the years 1914–15, Lloyd George passed
a note to Churchill which read, 'Had there been any other
Chancellor of the Exchequer your Naval Bill would have been
cut by millions.' Churchill scribbled a reply: 'There would also
have been another First Lord of the Admiralty! And who can
say . . . that there would not have been another government?'[25]

By the end of July 1914, however, these issues were suddenly
overshadowed by the extraordinary prospect of war between
the nations of Europe. The Cabinet met on a hot, sultry Friday
afternoon, 24 July, and spent several hours discussing the deadlock over Ulster. Then the Foreign Secretary, Sir Edward Grey,
was handed a note which he immediately read out to his
colleagues. It was the text of an ultimatum sent by Austria-
Hungary to Serbia, and it was clearly phrased so that no
self-respecting state could accept it. War suddenly looked likely.

But it took a while for the enormity of this document to sink in. Churchill later described how the border dispute in Ulster 'faded back into the mists and squalls of Ireland, and a strange light began immediately, but by perceptible gradations, to fall and grow upon the map of Europe'. He went straight to the Admiralty and immediately wrote down seventeen points which had to be attended to if war came. This piece of paper acted as a checklist for Admiralty officials over the next ten days.[26]

As the situation in the Balkans came to a head, the network of alliances that linked the nations of Europe divided the continent into two camps. If Austria declared war on Serbia, then Russia would come to Serbia's aid while Germany would support Austria. And if Germany went to war with Russia, France would come to Russia's aid. Knowing this, the German Army Command had devised a strategy known as the 'Schlieffen Plan'. This involved attacking France first and knocking it quickly out of the war, then turning to face Russia, which would be slower to mobilise. But the Schlieffen Plan necessitated passing across Belgian territory, which would draw Britain into the conflict because it had pledged to come to Belgium's aid in the event that it was invaded. As a result, within days, the whole of Europe slid helter-skelter into war.

Churchill was in his element. The Royal Navy's Home Fleet had gathered for its annual test mobilisation. The First Lord suggested that it should not disperse. Then he ordered the First Fleet to deploy from Portland on the south coast to the North Sea. This was tantamount to the fleet taking up its battle stations against Germany. Armed guards were put on naval supply depots and oil tanks.

At this moment, Clemmie was on holiday on the north Norfolk coast with their young family. Churchill, writing to her on 28 July, could not hide his excitement:

Everything tends towards catastrophe and collapse. I am interested, geared up and happy. Is it not horrible to be built like that? The preparations have a hideous fascination for me . . . We are putting the whole Navy into fighting trim . . . Everything is ready as it has never been before. And we are awake to the tip of our fingers. But war is the Unknown and the Unexpected . . . I feel sure however that if war comes we shall give them a good drubbing.[27]

On the evening of the following day, Germany declared war on Russia. Railways across Europe now started to move millions of men and thousands of tons of *matériel* in preparation for military activity. Churchill put the fleet on full mobilisation. Germany, as planned, prepared to launch its offensive against France and demanded right of passage through Belgium. Britain issued an ultimatum. On 4 August, as the German Army disregarded the ultimatum and crossed into Belgium, Britain declared war on Germany. At midnight, Churchill sent the order to all ships: 'Commence hostilities against Germany.' He rushed to 10 Downing Street where Lloyd George, who was already with Prime Minister Asquith, remembered his entrance thus: 'Winston dashed into the room, radiant, his face bright, his manner keen, one word pouring out on another how he was going to send telegrams to the Mediterranean, the North Sea and God knows where. You could see he was a really happy man.'[28]

Churchill was nearly forty. He had command of Britain's mighty navy at a time of major European war. And he was loving every minute of it.

2

Preparation: The War and the Wilderness

Churchill saw the war when it came in August 1914 as an opportunity to prove he could be a great war leader. But he had what might be called a roller coaster of a war. He started in high regard at the Admiralty. And by 1918 he was much respected as one of the leading organisers of the military success that brought victory on the Western Front. But in between he suffered probably the biggest humiliation of his life, and he was closely associated with a failure that haunted him for many years. So deep was his depression at one point in the war that he thought he might never recover from the unpopularity he had generated.

At the beginning of the war, Lord Kitchener was appointed by Asquith as Secretary for War. As we have seen, Churchill had previously criticised elements of Kitchener's command in the Sudan campaign of 1898, but now the two men, in charge of the two military departments of state, worked closely and effectively together. At this stage, there was no War Cabinet and the government of war continued without dramatic change from

the government of peace. Many people believed the war would
be a short business, that it would all be 'over by Christmas'.
Neither Kitchener nor Churchill shared this view. Kitchener
appealed for a million men to join the army and his face
adorned posters that went up across Britain. Hundreds of
thousands of volunteers patriotically came forward to do their
bit.

During the first few months of the war, the Royal Navy did
not perform as well as everyone, and especially Churchill, had
expected. Three elderly British cruisers were sunk by a German
U-boat while on patrol off Dogger Bank on 22 September with
the loss of nearly fifteen hundred men. And far away, along the
Pacific coast of Chile, a German battle squadron under the command of Count von Spee sank two further British cruisers with
similarly heavy loss of life. Before long, German surface vessels
and submarines, U-boats, also began to sink merchant ships in
large numbers. This was a challenge to the trade that Britain
relied upon for its imports. And the equipment of the Royal
Navy was found wanting. Mines sometimes failed to go off and
torpedoes went too low in the water, passing harmlessly under
the ship being targeted.

Churchill, however, soon committed himself to another campaign. The Germans had been turned back along the river
Marne, only a few miles from Paris. The Schlieffen Plan and the
strategy of knocking out France before attacking Russia had
failed, so the German Army sought out a defensive line in the
west. A 'race to the sea' began as both sides started to dig in and
construct a network of trenches. Churchill wanted to prevent
the Germans from occupying the Channel ports in Belgium,
and he visited Dunkirk and Antwerp to encourage their garrisons to hold out. The army was too stretched to provide extra
men for this and so Churchill committed naval troops and
Royal Marines, who also came under Admiralty command.

Bizarrely, for a First Lord of the Admiralty who should have been running naval affairs from London, Churchill spent several days in Antwerp trying to rally the defence of the city. He became obsessed with the defence of Antwerp and on 5 October he wrote one of the strangest letters of his life to the Prime Minister. He offered to resign from the Admiralty and the government to take field command in Antwerp as an army general supervising the city's defence. Fortunately, Asquith rejected the resignation, but Churchill still stayed in Antwerp for another four days. The city surrendered the day after he left.

The incident is revealing. Maybe Churchill was overexcited by the reality of war. Probably his yearning for military command got the better of him. Perhaps he really thought that he could make a difference and turn the course of the war at this crucial point. But his actions certainly displayed a strange lack of judgement and at his relatively young age show that he could be unhinged by military events. The Conservative opposition leaders seized on the incident, which they regarded as near farcical, and Bonar Law thought it showed that Churchill had become mentally unbalanced. Clementine, too, felt for some years that her husband's sense of proportion had deserted him at Antwerp.

Back at the Admiralty, Churchill was faced with the decision of whom to appoint as the new First Sea Lord to replace Battenberg, who resigned in the face of unpopularity over his German origins. (Of course, these origins were shared by the royal family, who wisely changed their surname to the very British-sounding 'Windsor' later in the war. Battenberg himself changed his family name to Mountbatten, and his son would be one of Churchill's leading generals thirty years later.) Churchill wanted someone aggressive in spirit and decided to bring back the seventy-four-year-old Lord Fisher. It was not a good decision. Fisher proved unpredictable, irascible, and like many senior people of his age, impatient and crotchety when he did not get his way. Churchill remained a great admirer of the elderly Sea Lord, but their disagreements mounted, and Fisher would have a great impact on the next phase of Churchill's career.

Another development at the Admiralty that had a lasting effect upon Churchill was the rapid progress in code-breaking. Before the war, Naval Intelligence had largely been preoccupied with intercepting enemy cables. But wireless telegraphy spread rapidly in the years before 1914, and the German Navy now had the ability to communicate with its ships at sea by radio in code. Many routine instructions and orders were sent daily by this method. In a lucky break early in the war, the Russians captured a German naval code book and passed it on to the British. The Australians then captured another code book, used by the Germans in communications with their merchant ships. In November 1914, when a third code book was found in a sunken German destroyer, the Admiralty had all it needed to decipher messages between ships at sea and their commanders at home. Churchill already knew the value of breaking enemy codes from his experiences in the Boer War and had forged links with the world of spies and espionage before 1914, so he was excited by this development. A new unit was set up known as Room 40, named simply after the room in the Admiralty where it was located.[1]

Churchill became obsessed with the control of the high-grade intelligence that was intercepted and interpreted by the cryptographers in Room 40. Most of the messages sent by the Germans made little sense in themselves, but they could provide vital information when properly analysed and placed into context by intelligence experts. Churchill resisted this and delighted in reading the transcripts 'raw', as they were sent. And he permitted very limited distribution of the intelligence outside the Admiralty building. Even the Cabinet was not regularly informed of the decrypts. This made for confusion and a succession of blunders in the first six months of the war. In December 1914, when orders were intercepted that a German naval raiding party was to cross the North Sea and shell mainland Britain, Churchill and his leading advisers at the Admiralty decided to order a naval squadron to intercept the German vessels on their return. The Germans shelled Scarborough, Hartlepool and Whitby, killing or injuring some five hundred civilians, but the Royal Navy squadron missed them on their return in the fog. The press were outraged and demanded to know why the most powerful navy in the world had allowed Britain to be shelled from the sea. 'Where was the navy?' asked the Scarborough coroner. Instead of a great triumph caused by the interception of highly useful intelligence, it turned out to be a low point for Churchill who was blamed for the national humiliation.

In January 1915, more ciphers detailing German naval movements in the North Sea were intercepted. This time the Royal Navy was waiting for the German vessels at Dogger Bank. It was the first time two navies equipped with great Dreadnoughts had clashed at sea. Churchill described the excitement of following this action on the charts on the Admiralty's walls:

> There can be few purely mental exercises charged with more excitement than to follow, almost from minute to minute, the phases of a great naval action from the silent rooms of the Admiralty . . . Telegram succeeds telegram at a few minutes' interval as they are picked up and decoded . . . and out of these a picture always flickering and changing rises in the mind, and imagination strikes out around it at every stage flashes of hope or fear.[2]

The *Blücher*, one of the German warships, was sunk, there were no significant British losses, the German High Seas Fleet withdrew, and a great victory was proclaimed. In reality, the Royal Navy could have caused far more damage, but the commander thought there were U-boats in the vicinity and failed to press home his attack. The Admiralty had not passed on to him the intercepted message that said the nearest U-boat was forty miles away and well out of the action.

For Churchill, knowledge was power, and he relished the fact that just he and a tiny number of senior officials around him knew what the Germans were saying to each other at any given time. After he left the Admiralty, Room 40 grew and its work improved, and lessons were learned about the best distribution of the intelligence gained. But overall, British Naval Intelligence had a good war, and Churchill was proud of his role in helping to pioneer this. In the 1920s he wrote, 'Our Intelligence service has won and deserved world-wide fame. More than perhaps any other Power, we were successful in the war in penetrating the intentions of the enemy.'[3] And Churchill retained a fascination with code-breaking and the use of what became known as signals intelligence (SIGINT). This would play an even more important role in the next war.

By 1915, it seemed to Churchill that the war had settled into a battle of attrition on land and stalemate at sea. Many millions of men faced each other in a line running from Switzerland to the Channel, now called the Western Front. Machine guns, heavily constructed defensive positions and huge masses of barbed wire prevented either side from advancing. At sea, the Royal Navy had successfully blockaded much of the German Navy in its fortified harbours, from which it did not dare venture out. Churchill argued that torpedoes and mines were to ships at sea what barbed wire and the machine gun were to soldiers trying to advance on land. His whirlwind mind threw out a variety of ideas and innovations to address these problems. They would have a lasting effect on the future face of war.

Churchill was particularly worried about the U-boat menace. At that time the German submarines would use their torpedoes only against the biggest ships but would come to the surface and fire their deck guns to force smaller merchant vessels to surrender. Churchill encouraged the use of decoy ships, or 'Q-ships', which looked like cargo vessels and flew the Red Ensign, the Merchant Navy's flag. However, when a U-boat surfaced, the Q-ship's sailors, a Royal Navy crew in disguise, threw open various trap doors and shutters to reveal their guns and then engaged the U-boat. Although they enjoyed limited success in sinking U-boats, the Q-ships were the subject of many boy's-own tales of derring-do.[4]

While still at the Admiralty, Churchill also began the process of developing 'land ships', giant armoured vehicles with cater-pillar tracks that could advance through barbed wire under fire. He suggested to the War Office that the army should develop armoured tractors that could break the deadlock of trench warfare. His ideas were rejected as being 'not likely to lead to success'.[5] So, in February 1915, Churchill committed Admiralty funding to the design of these strange vehicles. They were built in great secrecy and were disguised as 'water tanks for Russia', soon shortened to 'tanks'. Later, the army took this programme over and eventually, and still somewhat reluctantly, launched a new era in warfare with the use of the armoured tank at the Battle of the Somme in September 1916. Many people later laid claim to have invented the tank, but a Royal Commission credited Churchill with the 'receptivity, courage and driving force' that turned the idea into an effective instrument of war.[6] For his part, Churchill would never forget that the army had obstructed an idea which he thought could provide a war-winning machine.

He was also concerned about Zeppelins, giant airships from which the Germans could drop bombs on mainland Britain. Churchill wanted to find a way to strike at the Zeppelins so he authorised bombing raids on the Zeppelin bases in Germany. Primitive aircraft of the Royal Naval Air Service launched the first tentative bombing raids on the Zeppelin sheds. Thus, through his encouragement of the art of deception at sea, the development of the tank on land and the beginning of a very primitive form of aerial bombing, Churchill had shown himself very much in support of radical new ideas for warfare, and especially anything that he thought might break the stalemate along the Western Front. He later wrote that the soldiers, sailors, airmen, civilians and inventors who came up with these new ideas 'were a class apart, outside the currents of orthodox opinion, and for them was reserved the long and thankless struggle to convert authority and to procure action'.[7] Churchill was already beginning to identify the need for a War Lab of people and new ideas twenty-five years before he would muster his own when he found himself in charge of another war effort.

At the Admiralty, Churchill also sought out strategies by which Britain might take the offensive. He explored the option of attacking Germany's northern coast in the Baltic, from where an army could drive on Berlin. Then he focused on attacking Turkey, which had come into the war as Germany's ally at the end of October 1914. Churchill tried to whip up Cabinet enthusiasm for a naval attack in the Dardanelles, the narrow channel that joined the Mediterranean with the Black Sea. The Turkish capital, Constantinople (now Istanbul), was only a few miles further north. Churchill talked ambitiously about sending a naval task force to bombard the forts of the Dardanelles and an army group to land at Gallipoli and march north to knock Turkey out of the war in a single blow. Kitchener supported the plan, but it required a level of coordination between the army and the navy that was way ahead of its time. And not much thought had been given in pre-war planning to amphibious operations to land men on beaches.

After several months of debate, the Dardanelles offensive began with a naval bombardment on the morning of 18 March 1915. On the first day, three British and one French battle-ship hit mines and sank. It was a dreadful start. From this day on naval commanders concluded that they could never again enter the 'narrows' of the Dardanelles or penetrate as far as Constantinople. So, on 25 April, soldiers were landed on the beaches at Gallipoli. Many were Australians and New Zealanders, in a newly arrived force known as the Anzacs. From day one of the campaign things here went wrong also. The Turks had built up their defences over the previous month, and the men who waded ashore under heavy machine-gun fire were able to take only a narrow bridgehead of land. The Turks proved far more effective soldiers than had been anticipated, and despite much bravery and heroism, the whole Gallipoli campaign was soon mired in a similar stalemate to that which characterised the Western Front.

In mid-May, the campaign ushered in a major political crisis. Fisher, the First Sea Lord, had been proving more and more difficult to work with, at times supporting the Dardanelles campaign, then opposing it. He offered to resign on several occasions, but then always carried on regardless. Then on 15 May, he walked out in a huff. He refused to see Churchill and announced he had left for Scotland. Churchill was aghast that anyone could just walk out at a time like this. But this time the resignation was final. This might not have precipitated a polit-ical crisis had it not coincided with a scandal on the Western Front, where it was revealed that the army was suffering from a major shortage of shells. Questions were now being asked

about the ability of Asquith's government to manage the war.
During several days of crisis meetings it was resolved that
Asquith's Liberal government would be replaced by a Coalition
government. Leading Conservatives agreed to join the Cabinet,
but the price they demanded was the demotion of their old
enemy, Churchill. Lloyd George became Minister of Munitions
to sort out the shells crisis. Bonar Law, the Conservative leader,
came into government as Minister of the Colonies and his
fellow-Tory Arthur Balfour replaced Churchill at the
Admiralty. Churchill himself pleaded with Asquith, 'I will
accept any office – the lowest if you like that you care to offer
me.'[8] He was given the relatively humble post of Chancellor of
the Duchy of Lancaster, which he accepted because he could
remain in the War Council, albeit in a non-executive position.

Churchill was shattered by this turn of events. He felt he had
been made a scapegoat and was now unable to influence the
course of military events just when the war needed someone
with his abilities. The fighting in Gallipoli carried on until the
end of 1915, when an evacuation was organised. It cost 140,000
Allied casualties, many of them Anzacs. Conservative enemies
blamed Churchill for planning and orchestrating what became
known as the 'Dardanelles fiasco'. In the House of Commons,
they would shout 'Remember the Dardanelles!' when he got up
to speak. With his departure from the Admiralty many observers
thought that his political career was over. But there is no doubt
that the humiliation was a political matter. The Conservatives
had finally got their own back on this pushy, ambitious know-all
who had betrayed them ten years before. As far as Great War
military disasters were concerned, the Dardanelles campaign
was no worse than many others. But Churchill had been made
to pay the price for its failure. Although he remained in the War
Council, from now on no one listened to his arguments or propo-
sitions. He was pretty well ignored.

For the next six months, and for the first time since entering government in 1908, Churchill found himself under-employed. With time on his hands, and unable to exercise his great passion for military affairs, he took up painting, which brought him some sort of solace. His correspondence with Clementine comes almost to a halt during this period – largely because he was at home much of the time and they had no need to write to each other. She provided great support, and he still had Cabinet meetings to attend along with his regular parliamentary duties. For many people, that would have been more than enough. But for Churchill it was a period of immense frustration and, along with the sense of grievance he felt, provoked a profound depression which he called his 'black dog'.

In the autumn he decided he needed to get away from Westminster for a bit and the next remarkable phase of his extraordinary life unfolded. He resigned from the government and took up a commission in the army, fulfilling his wish to command men in wartime. In November, he arrived in France and spent ten days with the Grenadier Guards, learning for the first time the realities of life in the trenches. He wrote home: 'Filth and rubbish everywhere, graves built into the defences & scattered about promiscuously, feet and clothing breaking through the soil, water and muck on all sides . . . the venomous whining & whirring of the bullets which pass overhead.' But he concluded: 'I have found happiness and content such as I have not known for many months.' Later that month he wrote: 'I am very happy here. I did not know what release from care meant. It is a blessed peace.'[9] In January, he was appointed lieutenant-colonel and for the next five months the man who had been President of the Board of Trade, Home Secretary and First Lord of the Admiralty commanded eight hundred men of the 6th Battalion of the Royal Scots Fusiliers on the Western Front. A friend from the Liberal Party, Archibald Sinclair, was made his deputy. Churchill's new unit was made up mostly of Scottish volunteers who had answered Kitchener's appeal at the beginning of the war. Many of the men were ex-miners, while the officers had been young professionals. They were initially sceptical of their eccentric commanding officer, who wore strange clothes and regularly received hampers of cheese, ham and pies from home. But he showed great interest in the welfare of his men and quickly won their respect and admiration. Morale improved soon after Churchill and Sinclair took command.

For the months he served in the trenches, his unit was stationed in grim, semi-waterlogged trenches around Ploegsteert, near the Belgian border. It was a relatively quiet period on this stretch of the Western Front and casualties were not great. But it was still a dangerous place to be. And just as in his previous combat experiences, Churchill did not hide from danger. Far from it. He seemed almost reckless at times. He surveyed the trenches daily, sometimes from no man's land. And his letters home reveal several near misses. On one occasion, he had just left a dugout when a shell landed there, killing one officer and wounding several others. He was frequently showered with debris after shells burst near by. Once, a piece of shrapnel big enough to have taken his hand off landed two inches from his wrist.

However, this period of intense soldiering proved a great fillip to Churchill. His daily letters to Clementine from the trenches are particularly intense. He had time to reflect on his political career so far, at one point writing: 'My conviction that the greatest of my work is still to come is strong within me: & I ride reposefully along the gale.' But there was also time for regret and frustration. Once, having seen a German aircraft above the trenches, he wrote: 'There is no excuse for our not having command of the air. If they had given me control of this service when I left the Admiralty, we should have supremacy today.' When he heard of the progress being made with the development of tanks, he wrote: '[H]ow powerless I am! Are they not fools not to use my mind – or knaves to wait for its destruction by some flying splinter.' On another occasion, when a shell landed close by, he mused: '20 yards more to the left & no more tangles to unravel, nor more anxieties to face, no more hatreds & injustices to encounter; joy of all my foes, relief of that old rogue, a good ending to a chequered life, a final gift – unvalued – to an ungrateful country.'[10]

For her part, Clementine was worried sick each day that her husband spent in the trenches. 'I live from day to day in suspense and anguish,' she wrote. 'At night when I lie down I say to myself "Thank goodness he is still alive".' But she went on giving him support and good advice about his political future. Once, back home on leave, he made a particularly ill-judged speech in Parliament and Clementine wrote: 'To be great one's actions must be able to be understood by simple people.'[11] It was wise advice and she would come back to it again: Churchill had to think about how others would interpret his actions and his behaviour.

After nearly five months with the Royal Scots, Churchill's battalion was withdrawn from the front line and was due to be merged with another that had suffered heavy losses. The forty-one-year-old Churchill took this moment to leave the army and return to political life. He had done his bit of soldiering and could feel justifiably proud of his active service, even though he had never had to lead his men over the top in one of the assaults so typical of the Western Front. He had been a popular and devoted commanding officer, and the Royal Scots were genuinely sad to see him leave. Fifty years later, they would provide a guard at his funeral.

Soon after Churchill returned to Westminster, Field Marshal Douglas Haig, the commander-in-chief of British forces in

France, launched his 'Big Push' along the river Somme. Churchill was lucky to be out of it. Almost twenty thousand British soldiers were killed and forty thousand wounded on the first day alone. And the battle dragged on for nearly five months, leaving a death toll of British, French and Germans running into several hundred thousand. Churchill was deeply opposed to such futile head-on attacks without an overall superiority in men or guns. And he said so powerfully in the Commons and in the press.

In December 1916, Lloyd George replaced Asquith as Prime Minister in a reconstituted Coalition government. He tried to bring Churchill back into the Cabinet, but the Conservatives still objected. However, in early 1917, an inquiry cleared Churchill of any blame in the Dardanelles campaign and he began to behave more like his old self again. His speeches in Parliament once more displayed his talents and his understanding of the nature of the war. Churchill's standing picked up and in July 1917, Lloyd George finally won the argument with his Conservative colleagues and brought Churchill back into the government, in the key role of Minister of Munitions (although still without a place in the central War Cabinet). It was two years since he had been forced to leave the Admiralty. Probably the worst two years of his life.

The Ministry of Munitions was one of the vast new departments of state that had been created during the war. Three million workers were involved in producing and supplying munitions. Churchill, as ever, immersed himself in his new task of spurring on the armaments factories, in order to deliver the most efficient supply of weapons and shells to the front. He went back and forth on countless trips to France, to see his French equivalent and to set up joint munitions programmes, and to visit the front line. The scene of military activity still exerted a magnetic pull on him. He made it clear to Haig that he thought the Germans could not be defeated on the Western Front. But that didn't stop the British commander from launching another futile offensive at Ypres in the summer and autumn of 1917. By its finish, it had led to even more slaughter and the loss of half a million men killed, wounded or missing.

In early 1918, Churchill began planning for the build-up of huge numbers of tanks and aircraft that he hoped would be used to mount a new style of offensive in 1919. He visited France on many occasions and was staying in a billet near the front line on the night of 20 March when he was awoken at 4.30 a.m. by the sound of a massive artillery barrage. It was the beginning of Germany's great offensive in the west to try to win the war. A few weeks earlier, Russia had signed a peace treaty with Germany in the aftermath of the Bolshevik Revolution, so Germany was now free to rush troops from the Eastern to the Western Front. And Germany had to hurry, because the Americans, who had entered the conflict the previous year but so far had not deployed many men, would soon be arriving in great numbers to tip the balance in the Allies' favour. Within weeks, the German offensive had rolled the Allied line back several miles. The Germans recaptured in a matter of days all the territory that had so painfully been won from them along the Somme in 1916. Haig prepared to withdraw to the Channel ports.

Lloyd George sent Churchill back to France to assess the situation. Churchill found Haig's headquarters lacking any sense of bustle or excitement despite the fact that a hundred thousand British soldiers had been killed or captured. In Paris, he was invited by the Prime Minister Georges Clemenceau to visit the front with him. He was more impressed by the determination of the French commanders, especially General Pétain and Marshal Foch, who was appointed commander-in-chief on the Western Front. The line never broke. So the

Germans were prevented from turning a breakthrough into a breakout.

Back at the Ministry of Munitions, Churchill slept in his office and worked literally day and night to deliver the ammunition and weapons needed to replace the losses in France, pushing and cajoling industry to increase its output to the limit. By the end of April, the British Army had replacements for every gun, tank and aircraft lost so far during the German offensive. Haig wrote in his diary, 'He has certainly improved the output of the munitions factories very greatly, and is full of energy in trying to release men for the army and replace them by substitutes.'[12]

By the summer of 1918, the German offensive, which had come within artillery range of Paris, slowly began to run out of steam. Churchill shuttled back and forth to and from France, visiting command posts and gathering information about munitions requirements. On 8 August the British Army launched a counter-attack with two thousand artillery pieces and 456 tanks. German commanders spoke of this as the 'black day' for their army. The tide of war at last began to turn. Over the next month, the British and French pushed the Germans relentlessly back in retreat. The static war now became a mobile war. Tanks were used in giant mechanised thrusts to punch their way through enemy lines. Rolling artillery barrages moved forward just in front of the advancing troops. Aircraft flew endless sorties in support of the troops on the ground. By sheer force of arms, the huge British Army (at the time the biggest ever to be sent into battle), along with the French, Canadians, Australians, New Zealanders and Americans, defeated the German Army in the field. In early November, the Germans sought an armistice. On 10 November, Churchill attended a Cabinet meeting to discuss peace terms, and at 11 a.m. the following day the guns finally fell silent. Churchill was in his office alone and heard people assembling in the streets outside. He saw Trafalgar Square fill with cheering crowds. Clementine, who was heavily pregnant with their fourth child, joined him and they went to Downing Street to congratulate Lloyd George. Churchill later wrote, 'Victory had come after all the hazards and heart-breaks in an absolute and unlimited form ... All [the enemy's] armies and fleets were destroyed or subdued. In this Britain had borne a notable part, and done her best from first to last.'[13]

Churchill learned several lessons from the First World War that would prove invaluable when it came to leading the country in the Second. He reflected on many of these in his massive history of the war, *The World Crisis 1911–1918*, which he published in five volumes between 1923 and 1931.[14] Much of what he observed and wrote here became relevant in 1940. His suspicions of the limitations of admirals and generals hiding behind traditions and outdated custom and practice were confirmed. During the Boer War and the Great War, he felt that many military leaders lacked the necessary offensive spirit, and he would later be wary of this. Also, the failures he saw in the Great War confirmed his feeling that policy should clearly be set by politicians and then carried out by the military. He despaired at weak and indecisive political leadership and at any form of government organisation that did not allow clear, unambiguous decision-making to emerge. In wartime, political and military issues come together and Churchill felt that the Prime Minister should be intimately concerned with both setting and implementing military policy. He was also shocked by the fact that generals like Haig became such major public figures that they were virtually unremovable and could carry on with futile assaults despite opposition from their political masters. Churchill relished unorthodox thinking, whether inside the military or outside from civilian scientists or innovators. He thought the vast losses of the Great War were appalling, and avoidable. He would never rule over a government where human life was sacrificed so readily.

In the ten years that followed the end of the Great War, Churchill enjoyed a period of great activity, fame and considerable political success. This book is not the place to go through all the details, but some points have a bearing on his later story. In January 1919, Lloyd George appointed Churchill as Secretary for War and Air – a new position combining the traditional War Office responsibility for the army with responsibility for the growing field of aviation, both military and civil. This revived an enthusiasm in Churchill for flying, which he once again took up, having reluctantly given up lessons several years before. However, in July, he had a near-fatal accident at Croydon aerodrome just after taking off with his instructor. As his plane fell to earth Churchill thought, 'This is very like Death.' He was saved by his seat belt and was lucky to walk away from the wrecked plane, although his instructor was unconscious for some time. As before, Clementine and some of his friends pleaded with him to give up such a dangerous hobby. This time he did so for good, although his fascination with flying lasted for the rest of his life.

Churchill's first post-war political challenge was organising the process of demobilisation. Britain had nearly three and a half million men in arms at the end of the war and now most of them wanted to return home to resume civilian life as soon as possible. There was widespread discontent at how slowly this was being managed, and something resembling a mutiny took place near Calais. Haig suggested that the ringleaders should be shot. Churchill felt the whole issue was more in need of effective industrial-style organisation rather than military discipline and vetoed Haig's suggestion. Soon he came up with a fairer way to organise demobilisation, with the men who had been in military service the longest being the first to be demobbed, and the wounded given further priority. Within months, several million men had left the forces. The army rapidly made the transition to peacetime and a temporary boom in the economy helped men to find jobs when they got home.

A more tricky problem unfolded in Russia, where British troops were still stationed, supposedly to guard war supplies that had been sent there in 1917 and 1918, both during and after the revolution. Churchill was passionately opposed to communism, as were many members of the propertied classes in the West. He was also deeply upset by the treatment of the Russian aristocracy and especially of the deposed royal family, all of whom had been shot by the communists in 1918. Churchill was outraged at the atrocities committed by the Bolsheviks, who he said 'hop and caper like troops of ferocious baboons amid the ruins of cities and the corpses of their victims'. In contrast to many, he wanted to be magnanimous to the defeated Germans and to make war on the new communist regime, a policy he summed up as 'Kill the Bolshie, Kiss the Hun'.[15] In 1919, a civil war raged in Russia between so-called 'White' forces, who were opposed to the Bolsheviks, and the Red Army of the new regime. In London, the government wavered between committing British troops to the White cause and withdrawing its forces altogether. It was not an edifying spectacle and Churchill gained little credit from being in charge of the army at this time. At one point, Lloyd George told him to abandon his obsession with anti-communism, which, 'if you will forgive me for saying so, is upsetting your balance'.[16] Finally, as it became clear that the communist forces were about to defeat their opponents, British troops were ordered home. But Churchill would later be remembered by both the communist regime in Moscow and by the Left in Britain as the man who tried to strangle the Soviet Union at birth.

Much nearer to home, the problem of Ireland re-emerged, having been put under the carpet when war came. In 1921, the island was partitioned. The King opened a parliament for Northern Ireland and six of Ulster's nine counties retained their union with Britain. The status of the southern counties remained uncertain as all the elected Irish MPs were members of Sinn Féin who set up their own government in Dublin. Militants formed the Irish Republican Army and attacked the police and symbols of British rule in Ireland. In response, Churchill agreed to the establishment of the Auxiliaries to help maintain law and order. Better known as the Black and Tans, this mercenary group of ex-soldiers responded to terror with counter-terror and the situation degenerated into a vicious cycle of violence. Eventually, a truce was agreed; and in October 1921 two Irish leaders, Arthur Griffith and Michael Collins, led a delegation to London to negotiate the terms by which an independent state would be created. Churchill, who was by now Secretary for the Colonies, was one of the team of nego-tiators representing the British government. The negotiations went on for weeks. Churchill's total belief in the virtues of the British Empire made it difficult for him to comprehend why the Irish did not want to remain part of this great venture. And he was concerned that an independent Ireland might be a military backdoor that could be opened to Britain's enemies in a future war. The Irish delegation insisted that Ireland would remain neutral. Churchill was far from convinced. Despite their dif-ferences, however, Churchill took to Michael Collins, a soldier who had ordered acts of violence against the British. At one point, Collins told Churchill that there was a price on his head. Churchill responded by showing Collins a copy of the Boer poster that offered a reward for him, 'dead or alive'.

A treaty was finally agreed after a marathon session ended at three o'clock in the morning of 6 December. It fell to Churchill to sell the Irish Treaty to the House of Commons, where the majority of Conservatives regarded it as a betrayal of imperial rule. After Churchill gave one of his most brilliant parliamentary speeches, the Commons approved the treaty by 302 votes to 60. In Ireland, the treaty was rejected and a bloody civil war broke out, in which Michael Collins was killed. Churchill continued to work hard to negotiate a settlement between the Irish Free State and the Northern Ireland Unionists. His statesmanship did much to end this phase of the conflict and to restore his own reputation, partly healing the scar of the Dardanelles campaign.

As Secretary for the Colonies, Churchill also became involved in remapping the Middle East, a region that was becoming ever more significant in world affairs. At the outbreak of war, the area had been part of the Ottoman Empire. The British had made a series of promises to the Arabs in the Hejaz (now Saudi Arabia) that if they supported the Allies against the Turks, they would receive independence after the war. Simultaneously, British and French diplomats had negotiated a secret treaty in which they agreed to carve up the region after the war into their own spheres of influence. And in the Balfour Declaration of November 1917, Britain also promised that a Jewish 'national home' would be created in Palestine, as long as this did not harm the rights of the existing Arab population. These three conflicting promises were not only impossible to reconcile at the end of the war, but were to be the root of much trouble in the decades ahead.

In the last year of the war, Britain had enjoyed a considerable military victory when the army under General Allenby conquered the vast area of Palestine, Jordan and Syria. So, a British military administration was left in effective charge of much of the region after the war. Britain and France were granted various 'mandates' by the League of Nations to govern until such time as the local populations 'were able to stand alone'. Churchill was a great admirer of T.E. Lawrence, better known as Lawrence of Arabia, who had helped lead the Arabs in their revolt during the war. In March 1921, the two men led a substantial team on a tour of the region, and Churchill found himself playing the unlikely role of king-maker. At the Cairo Conference, he installed Emir Feisal, who had led the Arab revolt against the Turks alongside Lawrence, as ruler in Iraq. Moving on to Jerusalem, Churchill created a new entity to the east of the river Jordan called Transjordan and installed Feisal's brother, Abdullah, as ruler. (This royal dynasty still reigns in Jordan to this day.) While in Jerusalem, Churchill met separate delegations of Arabs and Jews. He overruled Arab objections to Jewish settlement on Palestinian land and wished the Zionists good luck in their quest to settle the land. He was excited by the idea of the Jews returning to Palestine and convinced that they would help develop the whole area. He would support the Zionist cause for the rest of his life and never appeared to understand Palestinian Arab objections to the loss of their land.

With the military budget pared right down, in these years Churchill helped the Royal Air Force to make the transition from war to peace. The RAF had been formed out of the alliance of the army's Royal Flying Corps and the Royal Navy's Air Service in 1918. But it struggled to find a role under Sir Hugh Trenchard. Churchill used it to quell potential uprisings in Somaliland and Iraq. For a fraction of the cost of sending in a large army troop, the RAF sent a few squadrons to bomb or machine-gun local rebels. The RAF policed sections of the Empire and survived, thankfully, to fight another day in very different circumstances.

In October 1922, Churchill fell ill and had his appendix removed. In those days, this was an operation that necessitated considerable rest and recuperation. While Churchill was recovering, Lloyd George's government fell and a general election was called. The electorate in Churchill's constituency of Dundee had changed dramatically following the granting of the vote to just about all men over twenty-one and women over thirty. As a tough working-class city, Dundee had also suffered in the economic downturn that followed the brief post-war recovery. Churchill could manage to visit his constituency only for the last few days of electioneering. He was in pain and unable to stand. He lost the election badly and went into a period of political exile. He had been a firm supporter of Lloyd George's coalition, but by now was not at all confident about his support for the Liberal Party, which seemed to be in terminal decline against the rise of the Labour Party. His defeat in the election prompted a witty comment that he was now left 'without an office, without a seat, without a party, and without an appendix'.[17]

In the hiatus that followed, Churchill for the second time in his political career changed sides. He returned to Parliament two years later as the Conservative MP for Epping in Essex, a constituency he would continue to represent for the next forty years. Maybe he had simply judged the age of the Liberal Party to be over, in which case he was proved right. He was also gently courted by the Conservatives, who felt his powerful oratory was needed on their side of the Commons. Added to this, his hostility to the socialism he believed the Labour Party would introduce encouraged him into the Conservative fold. After the third general election in as many years, in October 1924, Stanley Baldwin became Conservative Prime Minister and offered Churchill the position of Chancellor of the Exchequer. This was not only the second most important office of state, it was the post that Churchill's father had held nearly forty years before. He was thrilled, and served as Chancellor for the next five years.

Churchill's spell running the national economy is best remembered for his disastrous decision to return to the Gold Standard, which meant pegging the pound once again to the price of gold. 'The return to gold' was a mantra of 1920s politics and most advisers, including the Governor of the Bank of England and the mandarins at the Treasury, recommended it. The problem was that Churchill rejoined the Gold Standard at the pre-war level of parity. John Maynard Keynes, the finest economist of his generation, immediately denounced this decision and wrote a book entitled *The Economic Consequences of Mr Churchill*. He argued that sterling was now seriously overvalued, which would have dire effects on the economy by making exports overpriced. As a result, British industry would inevitably suffer. He correctly predicted that employers would try to cut wages and that unemployment would rise.

Churchill's five budgets as Chancellor are also interesting because he returned to the ideals of his reforming days by introducing a state-backed contributory pension scheme which would provide an income for men and women over the age of sixty-five. One recent historian has described this as a landmark piece of legislation affecting about fifteen million people and freeing many of them from dependence on the Poor Law.[18] On defence spending he adopted the Treasury line and introduced substantial cutbacks. (He conveniently forgot this when he called for rearmament ten years later.) Certainly, in the late 1920s, the prospect of war seemed remote. Germany was still recovering from its crushing military defeat and subsequent economic chaos. France and Italy were benign. The Soviet Union was inwardly focused. There were tensions in the Far East that prompted spending on a huge naval fortress in Singapore, but Churchill even trimmed the navy's budget. It would be several years before the Nazi Party began rearming Germany and posing a new threat to British security.

In May 1926, a long-standing dispute in the coal mines, where the owners wanted to reduce wages and increase hours, led to the nation's first and only general strike. Churchill was bullish in his opposition to the strikers throughout. He took charge of the *British Gazette*, the government's mouthpiece, which was still published while the rest of the press went on strike. For ten days he acted like a press baron, printing his own partisan views each morning. He also wanted the newly formed British Broadcasting Company (the forerunner of the British Broadcasting Corporation) to air the government line on the radio. Its general manager, John Reith, refused and insisted it must remain independent. But once the strike was over, Churchill took a conciliatory position, attacking the mine owners and supporting a minimum wage for the miners. Many of his men in the Scots Fusiliers had been miners and he felt for their plight.

During these busy and broadly contented years, Churchill's private and domestic life took on a form that would continue for many years to come. His income from book writing was considerable, and *The World Crisis* was well received. It has been described as a 'Mississippi of rhetoric, sweeping along with great narrative power though little interpretative depth'.[19] His journalism provided further steady income. These earnings, along with his ministerial salary, made him relatively rich. But he was profligate with his own, if not the nation's, money. He took long holidays in the south of France, where he regularly gambled and lost money in the casinos. He acquired a Rolls-Royce and employed a chauffeur. Meals were washed down with nothing but the finest champagne (vintage Pol Roger). He loved the finest Cuban cigars and smoked several every day. In 1922, he bought Chartwell Manor, a country house with about eighty acres of land, in a beautiful part of west Kent. It took two years to remove the dry rot and rebuild as Churchill wanted it

and this cost a small fortune, but in 1924 Winston, Clementine
and their four children were finally able to move in. (Their
daughter Marigold had died of septicaemia in 1921, leaving
Winston and Clementine distraught. Mary was born the
following year.) The staff grew to eight house servants and
three gardeners. Family life revolved around Winston and his
needs and amusements. The large drawing room, with superb
views across the Weald of Kent, became his pride and joy. Here,
he would write, study and pontificate. He conversed with
guests and held meetings with officials. Informal conferences
on international finance were conducted here. In the grounds,
Churchill drained ponds, built walls and gathered geese, swans
and pets. Even feeding the birds was carried out like a military
operation.[20]

At the dinner table, Churchill dominated conversation and
guests sat wrapped in a torrent of wit, entertainment, history,
politics and anecdote. Often it became a monologue. Only occasionally was Clemmie able to broaden the discussion to bring
others into the debate. Churchill presided over his own court,
and his courtiers became loyal friends for the next few decades.
Professor Frederick Lindemann, an odd and very snobbish
Oxford physicist, delighted Churchill by being able to reduce
complex scientific ideas to accessible soundbites. Brendan
Bracken, MP and businessman, amused Churchill with his witty
and indiscreet descriptions of political friends and enemies.
Lord Beaverbrook, the Canadian press baron, was a regular visitor and frequent commissioner of Churchill's words for the
Daily Express. Churchill admired Beaverbrook as a driven man
always doggedly determined to achieve what he set out to do.
Churchill respected people who stood up to him and he argued
furiously with some of his courtiers, especially Beaverbrook, but
he never held a grudge and everything was always made up
again in the morning. Clemmie put up with most of these

friends: she genuinely liked very few, and disliked many. Along with the courtiers, Churchill's circle included the Bonham Carters, the Duff Coopers, the Sinclairs and other social equals who delighted in being in the presence of such a witty and important figure.

Churchill's flamboyant, larger-than-life personality, his energetic intervention in the business of other Cabinet members, and his support for the welfare of the working man might well have brought him into conflict with the Conservative leadership before long, as it had in the Edwardian era. But in May 1929 Baldwin was defeated in the general election and Ramsay MacDonald's Labour Party came to power. Churchill feared that a socialist revolution would follow, destroying everything that was good about Britain and her Empire.

In reality Churchill was falling out of sync with the times. He was out of line with most Conservatives as well as the Labour Party. The leading example of this was his dogged opposition to any sort of self-government for India. Churchill was still very much of a Victorian in his mindset about India. He believed the 'white dominions' of Canada, Australia, New Zealand and (at this time) South Africa were shining examples of cultural, political and linguistic solidarity, beacons illustrating the virtues of the British Empire. He just could not accept that non-whites would be able to achieve the same level of government as white men. In this he undoubtedly maintained the racist views on which the imperial project had grown and flourished. Some of his other concerns about a self-ruling India are of more relevance to our post-imperial view today. He was worried that there would be violence between Hindus and Muslims; and he predicted that the 'untouchables', the lowest Hindu caste, would be a minority totally left out of Indian society and would suffer far more than they did under the Raj. But more than anything, he felt that ceding power in India, the 'jewel in the crown' of the British Empire, would be the beginning of the end of the imperial venture. In this he was probably right.

For five years, Churchill argued against what became the Government of India Bill. When it finally went through Parliament in 1935, he had virtually no supporters left on his side. He was offensive towards Indians, especially Gandhi, whom he called 'a seditious Middle Temple lawyer now posing as a fakir'. And he insulted many of his parliamentary colleagues, including MacDonald, Baldwin, Sir Samuel Hoare, and Lord Halifax, who had been Viceroy in India. Once again, contemporaries suggested that he was unbalanced, unreliable and out of touch with the times. He was increasingly seen as out of date and old-fashioned. As one biographer has put it, his message sounded 'cracked and tinny, like a record played on an Edwardian phonograph'.[21]

This partly explains why, after ten 'good' years mostly in government, he was left out of office for the next decade, which became what he called his 'wilderness years'. And the mood of politics changed from the largely optimistic years of the 1920s to the slump, poverty and divisions of the 1930s. Politicians from across the spectrum called for disarmament and appeasement while Churchill was still pounding his drum as a warlord, making bellicose calls for action.

Increasingly, he turned to the past. His literary output was immense throughout the 1930s. He dictated most of what he wrote and kept at least one and sometimes two secretaries busy from breakfast until well after dinner. He would pace up and down and the prose poured out of him. In 1930, he wrote more than forty features for the press on a wide range of political and social topics. In the same year, he published a delightful, intimate account of his younger days, *My Early Life*. In 1931, the fifth and final volume of *The World Crisis*, on the war in the east, was published; and the following year a series of essays entitled

Thoughts and Adventures appeared. Then he began a substantial undertaking, a four-volume history of his ancestor the Duke of Marlborough, which was published from 1933 to 1938. He also had time to produce a series of biographical essays called *Great Contemporaries* in 1937. And during these years he began an even bigger project. In many ways, his *History of the English-Speaking Peoples* came to sum up his view of the white British Empire, conjoined with a history of the United States. This project was so vast that it was not completed until the 1950s, although Churchill earned a lot from the initial advances and he was still working on the first phase of the book when he was interrupted by events in 1939. These books and the journalism were vital in sustaining the income Churchill needed to support his extravagant lifestyle, maintain his family and keep up Chartwell. But they also define his view of the past and show how this shaped his attitude towards the present and the future.

The view of history that emerges from the millions of words in these books is fascinating and is the foundation stone of Churchill's belief in who the British are and what their role in history had been. Churchill did not work like a professional historian. He often started to write (or, rather, dictate) long before he had done much research. He knew the general thrust of his argument. This greatly shocked Maurice Ashley, his research assistant for the Duke of Marlborough biography. But Churchill was always keen to get the facts correct. He reportedly said, 'Give me the facts, Ashley, and I will twist them the way I want to suit my argument.'[22] His books were always very readable: like his speeches, they were akin to free-flowing rivers, full of great stories and vivid characters all borne along by the driving current of narrative. But the central arguments that emerge in his history were like religious dogma to Churchill. He believed passionately in the virtues of the institutions of British government, monarchy and Parliament, supported by the pillars of the great aristocratic families. Along with this went his trust in liberty and freedom as interpreted by the English governing class. He was a great believer in the Whig interpretation of history: that is, England (more so than Britain) had a special destiny among nations. This concept was forged out of centuries of conflict between barons and kings, the violence of the Civil War and the political experiment of Cromwell's Commonwealth before everything came good in the Glorious Revolution of 1688, which brought in the perfect combination of a permanent Parliament and a docile monarchy. England had then taken on the burden of overseas conquests and had set out to defeat continental tyrants. This was where the Duke of Marlborough came in, destroying the ambitions of Louis XIV to rule the continent. Britain (as it had now become) next went on to build an empire that was the most just that had ever been seen, and its people enjoyed the richest and freest democracy the world had ever known. This account largely ignored the process of industrialisation and the progress of science. These were acknowledged as being part of the British genius but were not of particular interest to Churchill in his histories. As long as the nation's political balance was maintained, Britain had a duty to stand up to tyranny and military rule whenever it threatened personal liberty.

This romantic but also very political interpretation of history underscored most of Churchill's political ideas, from his views on India and Ireland to his confidence in the Empire. It was central to his opposition to Bolshevism and socialism. It was behind his view of how Britain related to Europe. And it explains his earlier hostility to the Kaiser and to an aggressive Germany. For a man to whom conventional worship meant very little, these beliefs pretty well became Churchill's religion. And they would be at the heart of his appeal to the nation in 1940.[23]

Churchill had the time to develop and express these ideas throughout the 1930s. Although he hoped to be invited back into government when Stanley Baldwin returned to power in 1935, his unpopular stance over India prevented this. And so, with this world view very much on his mind, Churchill began to dwell on the next issue that would come to dominate his thinking and would then take over his life and bring his career to a peak – the rise of Nazi Germany.

As part of the research for his biography of the Duke of Marlborough, Churchill had visited Germany in the summer of 1932 to tour the battlefield at Blenheim, where his ancestor had won his greatest victory. Even then Churchill had been alarmed by the 'bands of sturdy Teutonic youths' he saw marching by. A meeting was actually arranged with the leader of the right-wing National Socialist Workers' Party, Adolf Hitler, in a Munich hotel. But Hitler failed to turn up. The following year, when Hitler and the Nazis were elected to power, Churchill was not immediately concerned. However, he was deeply shocked when Hitler banned all other political parties, and particularly when he outlawed Germany's Jews from the apparatus of the state, national, local and municipal.

By 1936, Churchill had started to focus ever more of his attention on what he saw happening in Germany. He deplored the growing persecution of the Jews and the bully-boy tactics of the Nazi thugs who intimidated, rounded up and murdered political opponents. He was also deeply worried by German rearmament, about which he started to receive evidence from an unusual source. Churchill had met Desmond Morton on the Western Front in 1916, and Morton was now a neighbour, living only a few miles from Chartwell. He was director of the government's Industrial Intelligence Unit, and so able to study first-hand secret reports on Germany's rearmament, which went against the terms of the Treaty of Versailles. Morton leaked some of these reports to Churchill, who then used the information in various articles and speeches. There is still debate over whether Morton acted with official sanction or purely on his own initiative.[24] But he certainly became a regular member of the Chartwell court, and Churchill acquired reliable evidence of Hitler's flouting of the Versailles Treaty and Germany's burgeoning arms industry. He was the first senior statesman in Europe to sound the alarm about the growing threat posed by a resurgent Nazi Germany.

Churchill's warnings fell on deaf, or at least covered, ears for some time. The general sense in Britain was that the Great War really had been 'the war to end all wars'. Its memory was still recent and painful. Hundreds of thousands of men had been lost, their names filling long lists on war memorials in every parish and in schools and workplaces across the land. In addition, hundreds of thousands had been physically maimed or psychologically scarred. Huge numbers had been shocked by the futility of the conflict, as is evident in the many anti-war memoirs that came out from the late 1920s onwards. And thousands of pacifists joined groups like the League of Nations Union and the Peace Pledge Union to protest at any form of rearmament. The policies of appeasement and disarmament were genuinely popular among the vast majority of the population, viewed as the best ways to maintain peace.

Yet Churchill was in no sense a warmonger. He had no wish to see a call to arms (although his fascination with warlike subjects remained), and he had long held that the Treaty of Versailles was unfair on the Germans and that they had good reason to reject some of it. Many in Britain felt the same, although very few in France agreed with this. But Churchill was the first to see that Hitler was a demonic leader who had to be stopped, who had to be stood up to. He instinctively knew that appeasing the German Chancellor would not bring his demands to an end and ensure peace; rather, it would encourage him to demand more. The only option was to rearm in order to deter him. Churchill argued that preparation for war was the best way to maintain peace when dealing with a dictator like Hitler.

As events unfolded, Churchill's lone voice and his predictions of an imminent catastrophe began to make more sense. He was particularly concerned by the growth of the German Air Force, the Luftwaffe. Another whistleblower, Ralph Wigram from the Foreign Office, passed him information about how the German aircraft factories were dramatically increasing output. With the predominant view at the time being 'the bomber will always get through', Churchill conjured up a terrifying vision of thousands of incendiary bombs being dropped on British cities and millions of people being killed as a consequence. Hitler and his Nazi cronies protested and denounced Churchill as a warmonger. But in 1935 Baldwin invited him to join a secret sub-committee on Air Defence Research. At the first meeting he attended, Churchill heard about successful tests that were being carried out in the use of radio waves to detect aircraft (see Chapter 5). He was interested but had no idea how significant this invention of radar, as it was later called, would prove to be.

In March 1936, Hitler remilitarised the Rhineland, territory that had been freed of German troops under the terms of the Versailles Treaty. We now know this was a critical moment for the German leader. He gambled that the British and French would not try to prevent him, but he did not yet have an army powerful enough to carry through the reoccupation if opposed. As it was, the British and French governments allowed him to go ahead. With great prescience, Churchill warned the Commons that Hitler would soon be in a position to invade France through Belgium and Holland, and that Britain's security was further endangered.

Throughout 1936, more visitors came to Chartwell with information about the inadequacies in Britain's air defences or with news of military unpreparedness. Many were serving officers: Squadron Leader Charles Torr Anderson reported how little was being done to train RAF pilots; Brigadier Percy Hobart brought details of deficiencies in the army's tank programme. All of these stories of British military weakness were in contrast to what Churchill heard about Germany's military growth. In truth, Britain *had* started to rearm, slowly and quietly, and within very limited budget parameters; and designers and manufacturers were working on plans for the next generation of planes and other machines of war. But they would not come into service until the end of the decade. Churchill was convinced it was too little, too late. More needed to be done, quickly. And he continued to say so, repeatedly, in public.

People were finally beginning to listen when in the winter of 1936–7 Churchill dropped to another low point in his yo-yoing career. His natural enthusiasm for the monarchy led him to support King Edward VIII in the abdication crisis that suddenly erupted in December. Churchill badly misjudged the mood of the country and he was shouted down in the House of Commons. It was a humiliating moment. He thought his career was finished. His 'black dog' depression returned and sometimes he could not even sleep at night. He also had financial worries and seriously considered selling Chartwell. It was saved for him by the generosity of one of Brendan Bracken's rich banker friends.

In May 1937, Baldwin resigned and Neville Chamberlain became Prime Minister. He pledged to continue the policy of appeasement while encouraging further minor rearmament. There was no way he would readmit Churchill to government. In February 1938, Sir Anthony Eden resigned as Foreign Secretary because he could no longer support the policy of

appeasement. Churchill admired him for this but thought the
future now looked grim. He told the Commons, 'I predict that
the day will come when, at some point or other, you will have
to make a stand, and I pray to God, when that day comes, that
we may not find through an unwise policy, we have to make
that stand alone.'[25]

In March 1938, Hitler announced the Anschluss, effectively
a German takeover of Austria. Within twenty-four hours, thousands of Austrians who opposed Nazi rule were imprisoned
or shot by the Gestapo. Churchill condemned Germany in the
House of Commons and called for an alliance against further
German aggression. This time many people rallied behind
him. Harold Nicolson MP wrote in his diary, 'Winston makes
the speech of his life'.[26] But, once again, the governments in
London and Paris did nothing.

By now, the Luftwaffe had become far bigger than the RAF,
and the Air Staff felt it could no longer guarantee the defence of
Britain. But the Cabinet rejected plans to increase spending on
RAF defences. Now it was the turn of the Minister for the Air,
Lord Swinton, to resign. Churchill despaired further that his
warnings were not being listened to.

During the summer of 1938, Hitler started demanding that the
Sudetenland region of Czechoslovakia be incorporated into
the growing German Reich. The area had a substantial Czech-
German population. This prompted a major diplomatic crisis.
Yet again the British and French governments decided they
must let Hitler have his way. They did not consult the Czecho-
slovak government on the matter. In September, Prime Minister
Chamberlain shuttled back and forth between Britain and
Germany. Along with the French Foreign Minister, Edouard
Daladier, Chamberlain ceded every point to Hitler. They believed
him when he said that acquiring the Sudetenland was the end
of his territorial ambitions. In Munich, at the end of a series of

meetings, Chamberlain persuaded Hitler to sign a piece of paper which stated that he had no wish to make war on Britain. The Prime Minister returned to London a hero. At Croydon airport, he waved the piece of paper to cheering crowds. He said he had brought back 'peace in our time'.

Churchill took a different view. In Parliament, he told MPs:

> [W]e have sustained a defeat without a war . . . we have passed an awful milestone in our history, when the whole equilibrium of Europe has been deranged . . . do not suppose that this is the end. This is only the beginning of the reckoning. This is only the first sip, the first foretaste of a bitter cup which will be proffered to us year by year unless, by a supreme recovery of moral health and martial vigour, we rise again and take our stand for freedom as in olden time.[27]

In response, Chamberlain merely derided Churchill for his lack of judgement.

In March 1939, in blatant disregard of everything he had told Chamberlain six months earlier, Hitler's army occupied the rest of Czechoslovakia. Chamberlain's reputation and the Anglo-French policy of appeasement were now in tatters. Churchill had been proved right and the tide of events was at last beginning to flow in his direction. He seemed to be the right man to stand up to Hitler. As the government hastily drew up a treaty promising to defend Poland's integrity, a 'Bring Back Churchill' campaign started in the press. Even the Labour Party came round to thinking Churchill should be back in power.

Churchill didn't have long to wait. Towards the end of August, Germany signed a surprise pact with Stalin's Russia. The ideological enemies had become military allies. The way was clear for Hitler's next act. The House of Commons was recalled from its summer recess. On the last day of the month, Churchill was still immersed in his history writing. He confided to a friend: 'It is a relief in times like these to be able to escape into other centuries.'[28]

Events at the beginning of September 1939 brought to a climax the many years of appeasement. Churchill's dire predictions during his wilderness years had come to pass. On Friday the first of September the German Army invaded Poland. The British and French governments both issued ultimatums to Berlin that German forces must withdraw. They had finally drawn a line in the sand after years of letting Hitler get his way. Chamberlain at last recalled Churchill into the government. But then the Prime Minister hesitated. Churchill waited for more than twenty-four hours not knowing what position he would be offered.

At 11.15 a.m. on 3 September, Chamberlain made a radio broadcast to the nation from the Cabinet Room in 10 Downing Street. He told the British people that Hitler had ignored the government's ultimatum, and that Britain was 'once again at war with Germany'. The broadcast featured in many documentary films at the time and has appeared in countless television programmes since.[29] It powerfully captures the drama of the moment.

Later that day, Chamberlain formally offered Churchill the post of First Lord of the Admiralty in the newly formed War Cabinet. It was a moment of triumph for the sixty-four-year-old Churchill, who had spent more than ten years out of office and many thought had come to the end of his political career. Returning to government in any capacity would have been exciting, but returning to the Admiralty and the position he had very happily occupied between 1911 and 1915 was a special joy. When Churchill entered the Admiralty building at about six o'clock that evening, he went straight to the office he had worked in when war was declared in 1914. With its dark oak panelling, grand Grinling Gibbons carvings over the fireplace and large portrait of Nelson, it had changed little. Churchill asked, 'Where is the octagonal table?' His old table was soon produced. He also asked for the chart box he had used before, and the maps were soon hung up. 'These are the same charts I used in 1915!' he exclaimed with evident glee to the officer who would be his naval assistant.[30] Churchill's appointment was for him just reward for years of work calling upon the nation to be prepared for inevitable war with Nazi Germany. But in the mood created by the declaration of war it was a genuinely popular appointment. That evening a message went out to the fleet: 'Winston is Back'.[31]

The Second World War had begun, but nothing very much happened in Britain. At the time, the first few months were called the Bore War, and they later became known as the Phoney War. Most of the children who had been hastily evacuated over the weekend that war was declared returned home to anxious parents before the month was out. All of the action was taking place on the continent. The German Army invaded Poland from the west and succeeded in conquering the country in less than a month. At the same time, the Red Army invaded from the east. Poland ceased to exist. Hitler committed all of his armoured panzer units and most of his army to Poland, but the French, politically divided and worried about German reprisals, did nothing. They sat behind the mighty defensive wall known as the 'Maginot Line', which ran along France's border with Germany. One British general described it as a 'Battleship built on land'.[32] Meanwhile, the British assembled a small expeditionary force of nine divisions that crossed to the continent in October.

From his office in the Admiralty, Churchill threw himself into war planning. The Royal Navy was still a powerful force in

1939. There was a lot to command and a lot to organise. Once again, he faced a major challenge. He settled into a way of working that would become his style for the next few years of war. He kept everyone on their toes with a barrage of minutes or memos – dictated messages sent to his advisers and colleagues asking for information, cajoling, upbraiding or requesting action. Some were witty, some solemn. Like emails today, they could be misinterpreted by their recipients, as being more fierce or stern than they were intended to be. He sent several of these minutes every day, hundreds every month. Some related to high-level matters of strategy, such as how to deal with the German U-boat threat, or ideas for brand-new initiatives that needed to be explored. Others concerned the smallest details, like the specifics of dealing with mines. Today, Churchill would be known as a 'micro-manager'. No detail was too small to escape his attention or to avoid his fury if not attended to in a way he considered fit. This led to immense frustration and exhaustion among those who worked closely with him. But the Royal Navy, Britain's oldest and in many ways most conservative military service, was galvanised by the presence of this human dynamo in the Admiralty. One of his naval aides later described how 'He practically killed people by overwork, and at the same time inspired people to extreme devotion.' Kathleen Hill, who had joined him at Chartwell to take down dictation, went with him into government. She later recalled: 'When Winston was at the Admiralty, the place was buzzing with atmosphere, with electricity. When he was away, on tour, it was dead, dead, dead.'[33]

There were many issues to concern Churchill over the first few months of war. Getting the navy on to a war footing was at the top of the 'to-do' list. German undersea telephone cables to the outside world were cut on the very first day. A blockade of the German fleet had to be organised and enforced. And

Churchill insisted on the need to equip the Royal Navy with
radar, which had not been done before the war as a cost-cutting
measure.

Just like in 1914, things did not all go well. Part of the fleet
was assembled at Scapa Flow, the vast anchorage in Orkney
that had been used in the First World War. From here, the
fleet could move out to intercept the German Navy if it tried
to venture into the Atlantic. Prevailing wisdom was that this
anchorage, the holiest of holies for the Royal Navy, was totally
secure. But in the early hours of the morning of 14 October,
a dark night with high tides and little moonlight, a U-boat
managed to sail on the surface of a small channel, past the anti-
submarine defences into Scapa Flow. Its commander looked
around for the best target and fired four torpedoes at a huge
First World War-era Dreadnought at anchorage that night. HMS
Royal Oak went down in just eleven minutes and 810 men were
lost, including an admiral and, horrifically, 150 boy sailors aged
fifteen or sixteen who were sleeping on board. In the chaos that
followed, the U-boat slipped back out of Scapa and returned
to Germany. Its captain was fêted by Hitler in a triumphant
parade down Unter den Linden in Berlin and he became one
of the first German heroes of the war, launching the legend of
the tough, daring U-boat menace that would haunt Churchill
many times over the following years.

Churchill was always moved by losses at sea, and when he
was told of the sinking of the *Royal Oak* it is reported that tears
sprang to his eyes. As he wept he muttered, 'Poor fellows, poor
fellows trapped in those black depths.'[34] The loss of this great
old battleship from another era was a human tragedy, but it did
not seriously affect the capability of the Royal Navy. However,
it did lead to the repositioning of the fleet to Rosyth on the Firth
of Forth. And typical of his concern for detail, Churchill followed every stage of the design and construction of the new sea

defences that he ordered to be built at Scapa Flow. Fittingly, these later became known as the 'Churchill Barriers'.

The loss of the *Royal Oak* and concern about aerial reconnaissance photography of ships in anchorage or at sea prompted Churchill to look for an imaginative way of deceiving the enemy. This was the sort of challenge he relished. He called for the building of 'dummy' ships, constructed out of wood and canvas, which would look just like the real thing in the aerial photos taken by the enemy. He had encouraged something similar when First Lord during the First World War. Again, Churchill was obsessed with detail. When he observed one of the dummy ships on a later visit to Scapa he said no one would fall for it because there were no gulls flying above it, whereas real ships were always surrounded by seabirds. He ordered that scraps of food should be put out of the front and behind to ensure that gulls constantly circled the dummy ships.

In October, having defeated Poland, Hitler offered peace terms to Britain and France. Churchill argued forcefully to his colleagues in the War Cabinet to reject the approach. The *Daily Mirror* said it was Churchill's 'brilliant memorandum' that stiffened Chamberlain's proposed reply to Hitler. He also wanted the RAF to bomb the Ruhr industrial area, but this time he lost the argument in the War Cabinet. Some of his colleagues felt that such an action risked provoking Nazi reprisal bombings on England. The Secretary of State for Air even argued that bombing the Ruhr would not be appropriate because it was private property![35]

As the months passed, Churchill grew hungry for action, and increasingly frustrated at the negative and passive views in the rest of the War Cabinet. He came up with a plan to drop mines in the Rhine to destroy shipping in one of Germany's major transport arteries. This idea was opposed first by his Cabinet colleagues and later by the French. Then he spent much time developing a plan to attack Norway and Sweden in order to prevent Swedish iron ore from being exported to Germany. This would have struck a mortal blow to the German war economy. But the prevailing view was that Britain could not assault the integrity of neutral Norway and Sweden. Such lack of offensive spirit among his Cabinet colleagues, many of whom of course had been the leading appeasers throughout the 1930s, began to frustrate Churchill more and more.

After intense debate over many months, he finally persuaded the Cabinet of the need to mount a campaign against Norway. The date was repeatedly put back but was at last scheduled for mid-April 1940. The navy would mine Norwegian waters against German shipping and the army would seize the port of Narvik, from which much Swedish iron ore was exported. But Hitler beat Chamberlain's government to it. On 9 April he launched a daring invasion of Denmark and Norway in a combined naval and air operation, dropping airborne troops to capture key airfields – the first time this had been attempted.

The British and French governments were taken entirely by surprise. The earlier plan to land in the north at Narvik was switched at the last minute to a landing in Trondheim. This was then changed again to a pincer movement to the north and south of the city. Troops were landed hastily without artillery, anti-aircraft weapons and other essential equipment. The navy also failed to coordinate with the RAF, and in early May a rapid evacuation was ordered. The whole campaign was a horrible reminder of the fiasco in the Dardanelles in 1915. The Chief of the Imperial General Staff, General Edmund Ironside, chafed at the lack of clear direction of the campaign by his political masters and at the muddle that ensued. 'Always too late. Changing plans and nobody directing,' he wrote in his diary. '[V]ery upset at the thought of our incompetence.'[36] These were tough words from the nation's top soldier.

The Royal Navy lost several ships, including the aircraft carrier HMS *Glorious*, although the Germans also suffered major naval losses, including the heavy cruiser *Blücher*. Churchill was infuriated by the whole episode. He felt it revealed that the British war effort was badly led and lacked any coordinating body powerful enough to make key decisions and then ensure they were followed through. And he was not alone. A two-day debate in the House of Commons was turned by the Labour Opposition into a vote of censure. Many Tories turned against their own leaders, too. Leo Amery, an ex-Cabinet minister, pointed to the front bench and quoted the words Oliver Cromwell had once uttered in Parliament by saying: 'Depart, I say, and let us have done with you. In the name of God, go!'

Churchill was in a difficult position. He had argued for a campaign against Norway in the War Cabinet for six months. But as First Lord of the Admiralty he had overall responsibility for a campaign in which the Royal Navy had hardly excelled. He could suffer by being closely associated with its failure. But how could he denounce a government of which he was a senior member? He knew that he stood a good chance of taking over from the discredited Chamberlain. Churchill was associated in many people's minds with pursuing a more rigorous war policy. Chamberlain was still associated with the failed policies of the 1930s. Churchill had already received several approaches and suggestions that he should take over the reins of power.[37] On the other hand, if the government fell, he might fall with it. How could he help to bring down the government without bringing himself down with it?

At about ten o'clock on the evening of 8 May, Churchill rose to speak in the Commons to wind up the censure debate. He put on a fine parliamentary performance. Henry Channon MP wrote, 'One saw at once that he was in a bellicose mood, alive and enjoying himself, relishing the ironical position in which he found himself.'[38] Churchill did not betray Chamberlain, and he took full responsibility for the poor performance of the Royal Navy off the coast of Norway. But in a final blaze of oratory, he declared, 'let pre-war feuds die: let personal quarrels be forgotten, and let us keep our hatreds for the common enemy . . . At no time in the last war were we in greater peril than we are now and I urge the House strongly to deal with these matters.'[39] In the vote that followed, thirty-three Conservatives and other supporters voted with the Opposition and sixty abstained. Chamberlain's majority of 213 was reduced to 81. As the Prime Minister left the chamber, howls of 'Go! Go! Go! Go!' were directed at him.

Confronted by this humiliation, the following day Chamberlain decided that it was time to form a 'national government'. It was impossible to consider going to the people in a general election with the war at this critical stage. So the Prime Minister asked the Labour Party if they would join his government. Clement Attlee, the Labour leader, consulted with his colleagues, then made it clear that they would not join a Cabinet led by Chamberlain. So the question became: who *could* lead a government that would command the support of the Labour Party and the whole nation?

Later that day, Chamberlain called Lord Halifax, the Foreign Secretary, and Churchill to a meeting at 10 Downing Street. David Margesson, the Conservative Chief Whip, was also present. Halifax was Chamberlain's natural successor. He was urbane, aristocratic, an eminent Yorkshire landowner, maybe a little aloof, but many thought a natural leader. He was also admired for his sound judgement. Whereas Churchill was thought by many to be impetuous, untrustworthy, unpredictable and difficult to restrain. Furthermore, Halifax had the support of the King. However, Churchill was the one most associated with fighting the war rigorously.

According to Churchill, Chamberlain asked the two men whom he should recommend to the King to replace him after his own resignation. The King was almost certain to follow the advice of his departing Prime Minister. Many years later, in his wartime history, Churchill described what happened next:

> I have had many important interviews in my pubic life, and this was certainly the most important. Usually I talk a great deal, but on this occasion I was silent ... As I remained silent a very long pause ensued. It certainly seemed longer than the two minutes which one observes in the commemoration of the Armistice.[40]

John Colville, later his private secretary, said that Churchill suspected a trap. If he proposed himself, he might be seen as too pushy. If he said he thought a peer could become Prime Minister, then Chamberlain might go for Halifax. So he stood with his back to Chamberlain, gazed out of the window towards Horse Guards Parade, and held his counsel.[41]

Eventually, according to Churchill, Halifax was the first to speak. He said that he felt he could not lead the nation in war from the House of Lords, that he would be in a hopeless position. He suggested Winston was a better choice. Churchill did not demur. Chamberlain took this as agreement to recommend the King should ask Churchill to become Prime Minister.

Churchill's account was written seven years after the event, and while making a splendid story out of the meeting no doubt has benefited from being recounted a few times. And it gets a few core fundamentals wrong, like the date of the meeting and who was there. On the other hand, Halifax's diary account of the same meeting, which was written that evening, makes no mention of any long pause. Halifax says that he felt an ache in his stomach at the thought of becoming Prime Minister at such a critical moment and so ruled himself out. He then claims that Churchill, 'with suitable expressions of regard and humility', eagerly accepted that he was the right man for the job.[42]

But that was still not the end of the matter. Chamberlain hesitated before going to the Palace to resign, so Churchill went back to the Admiralty. That evening, he had dinner with a few close colleagues and was described as being 'quiet and calm'. Later that night, his son Randolph telephoned from his army base to ask what was the latest news. Churchill replied, 'I think I shall be Prime Minister tomorrow.'[43]

3

Action This Day

What a difference a day makes. At 5.30 a.m. on 10 May 1940, Churchill was awoken in Admiralty House and told the news that German forces had launched an invasion of Holland and Belgium. Hitler's long-awaited and much-predicted attack upon the democracies of Western Europe had begun. The Phoney War was over. The real war was on.

Churchill was thrown into a round of urgent meetings as telegrams and messages poured in. The huge French Army was still stationed along the Maginot Line, the string of immensely powerful fortresses constructed along France's 250-mile border with Germany. Hitler simply bypassed this and threw his armies at neutral Holland and Belgium, bombing airfields and using airborne troops to capture key defensive positions. For the first few days, this was where the action was.

At eight that morning the War Cabinet met and authorised the sending of British troops to assist the Dutch and the Belgians. An hour later, Churchill was back at his desk in the Admiralty. The BBC had been broadcasting news of Hitler's invasion since seven o'clock and by now the entire country knew that a full-scale war crisis was on. But Chamberlain, having decided the previous day that a national government must be created, wavered again. He spoke with friends and said that maybe a change of government should be delayed until the battle on the continent was over, perhaps now was not the time for a change of leadership in Britain. Wiser colleagues, like Sir Kingsley Wood, a senior figure in the Conservative Party, spoke with Chamberlain and insisted the crisis made it even more important that he form a national government. Still Chamberlain hesitated.

The War Cabinet met again at 11.30 a.m. Ministers were beginning to feel that if a change of government were coming, it should happen sooner rather than later. But the overwhelming military business of the day crowded out any political debate. The news from Belgium was not good. Reports suggested that German airborne troops had seized a key fortress in a daring dawn raid.

In the late afternoon, the War Cabinet met for the third time that day. It heard that German infantry and tanks were now pouring into Belgium near Liège. Finally, the Prime Minister spoke up and told members that he intended to resign that evening and that a national government should be formed urgently. Chamberlain had at last accepted the inevitable and was about to bow out, despite the alarming news from across the Channel. He did not tell the Cabinet whom he would be recommending as his successor.

In accordance with tradition, Chamberlain went straight to Buckingham Palace to hand in his resignation to George VI. The King expressed his preference for Halifax as 'the obvious man' to become Prime Minister. When Chamberlain said Halifax had ruled himself out, the King knew that only 'one person . . . had the confidence of the country, & that was Winston'. Chamberlain agreed and left.

At about 6 p.m., Churchill was summoned to meet the King. No crowds had gathered as everyone was preoccupied with the news from Belgium, so Churchill slipped quietly into the Palace. According to his account, at this supreme moment in the nation's history, the King joked for a moment and asked, 'I suppose you don't know why I have sent for you.' Churchill responded wryly, 'Sir, I simply couldn't imagine why.' After a brief discussion, the formalities were completed and in a matter of minutes the meeting was over. In his diary, the King noted, 'He was full of fire & determination.'[1]

Churchill returned to the Admiralty as Prime Minister. The nation was in crisis. Less than three hundred miles away two million men were engaged in a furious battle. Events were moving at a stunning speed. Within weeks, Britain would be standing alone to face one of the most determined and ferocious enemies in its history. The burden on the new Prime Minister stepping up to lead the nation at this critical point was immense. Churchill must have felt doubts about what lay ahead. He confided to his detective in the car on the way back from the Palace: 'I hope it is not too late.' But he later wrote that he felt a profound sense of relief, that at last he had the authority to direct the entire operation of the war effort. In his Second World War history he wrote momentously, 'I felt as if I were walking with destiny, and that my past life had been but a preparation for this hour and for this trial.'[2]

At nine that evening Chamberlain broadcast to the nation, explaining that recent events made it clear that a national coalition government was needed. He told the British people that he had resigned and that Winston Churchill was their new Prime Minister. He asked everyone to give Churchill their full support. In Barrow-in-Furness, housewife Nella Last was keeping a diary for Mass Observation. In it she wrote that Churchill was a popular figure among the shipyard workers of the town and reflected that if she had to spend her 'whole life with a man, I'd choose Mr Chamberlain, but I think I would sooner have Mr Churchill if there was a storm and I was shipwrecked'.[3] The storm had arrived.

Churchill immediately started putting together his national government. And he set about restructuring the war effort to streamline the complex and cumbersome decision-making system that had been in existence up to this point. Later that same evening, he sent the King a letter outlining his top five appointments to the War Cabinet with Labour, Liberal and Conservative members in key positions. His government would be a true coalition, representing all political view-points across the nation. There was no vindictiveness towards Chamberlain, the man who had derided Churchill's views about the inevitability of war for years, who became Lord President of the Council. But this was not just an act of gen-erosity: Chamberlain was still leader of the Conservative Party and Churchill knew that he could not govern without his sup-port. Looking back at the end of a dramatic day, Churchill later wrote: 'I thought I knew a good deal about it all, and I was sure I would not fail. Therefore, although impatient for the morning, I slept soundly and had no need for cheering dreams. Facts are better than dreams.'[4]

It's difficult, knowing what we do today about events of the next few months and years, to realise that Churchill was by no means a universal choice as leader in May 1940. Officials in Whitehall were not at all heartened by his appointment. John Colville, one of Chamberlain's private secretaries at the time, remembers 'the mere thought of Churchill as Prime Minister sent a cold chill down the spines of the staff at 10 Downing Street ... His verbosity and restlessness made unnecessary work, prevented real planning and caused friction ... Our feel-ings ... were widely shared in the Cabinet offices, the Treasury and throughout Whitehall.'[5] This might be the sentiment of civil servants who feared change at a time of national crisis, but there was widespread suspicion of Churchill. He was thought to be impetuous, hot-headed and interfering. Even Halifax felt that he was led by his emotions rather than by reason. In British politics every Prime Minister has to have a substantial party base. But again, here Churchill was weak. Many Tories did not feel at all comfortable with him and had never forgiven him for deserting the party for the Liberals over thirty years earlier. He was seen as an opportunist and a chancer. Many felt like Lord Davidson that they simply 'don't trust Winston'. Old criticisms of him died hard. Alfred Lyttelton had said of him many decades before, 'He trims his sails to every passing wind.' Many Conservatives were also suspicious of his policies. He had supported Lloyd George's radical reforms of the Edwardian era, which had introduced unemployment benefits and state pensions. He had been a controversial Chancellor of the Exchequer in the 1920s. He had isolated himself over policy towards self-government in India. And, although his predictions about Hitler and Nazi aggression had been proved right, he had not won many friends in the Conservative Party, which had continued to support the policy of appeasement through the 1930s. Meanwhile, although the Labour Party now supported him as leader of a coalition government, this was more out of opposition to Chamberlain than active support for Churchill. And he was not a friend of the Left. He was still remembered for having threatened the miners at Tonypandy. Some newspapers, such as the *Daily Mirror* and the *Daily Mail*, were pro-Churchill. But the establishment paper, *The Times*, was unhesitatingly anti.

Without doubt, had he wanted to, Halifax could have become Prime Minister in May 1940. But it seems he saw the future as an impossible challenge, what might be described today as a 'no-win' situation. Churchill became Prime Minister partly by default, but also because his background, his love of military affairs and his sense of history made him (and others) feel that this was his moment, that his time had come. And he never lacked the confidence that he could govern and drive the country forwards to victory. In his first speech to Parliament as Prime Minister on 13 May, he repeated what he had said to his Cabinet colleagues earlier in the day, 'I have nothing to offer but blood, toil, tears and sweat.' Then he went on to give his first great speech as a wartime leader. As Hitler's panzers charged through Belgium, apparently unstoppable, Churchill told the House of Commons:

> We have before us an ordeal of the most grievous kind. We have before us many, many long months of struggle and of suffering. You ask, what is our policy? I will say: It is to wage war, by sea, land and air, with all our might and with all the strength that God can give us; to wage war against a monstrous tyranny, never surpassed in the dark, lamentable catalogue of human crime. That is our policy. You ask, what is our aim? I can answer in one word: It is victory, victory at all costs, victory in spite of all terror, victory, however long and hard the road may be; for without victory, there is no survival. Let that be realised. No survival for the British Empire, no survival for all that the British Empire has stood for . . . But I take up my task with buoyancy and with hope . . . and I say, 'Come then, let us go forward together with our united strength.'[6]

These were grand words and noble sentiments, but in the face of the rapid advance of Hitler's forces in Northern Europe they must have sounded hollow to those who heard them. How on earth could Churchill deliver?

Of the many tasks Churchill faced on becoming Prime Minister, first was the need to assemble his government. This was done while the war raged across the Channel. Some key members of what would become the War Lab were now put in place. At the centre of government, Churchill appointed himself Minister of Defence as well as Prime Minister. There was no precedent for this. It gave him unique authority to oversee all matters of military policy. The joint role put him in total control of Britain's war effort. There was virtually no aspect of war administration he could not influence. And this would be a vital feature of his leadership. There had been no such position as Minister of Defence in the First World War, and no ministry of this name during the 1930s. And, unlike from the 1960s onwards, no great department of state would report to him as Minister of Defence. The key adviser he appointed was General Sir Hastings Ismay, who would lead a small staff of a dozen officers and act like a staff officer, representing Churchill at Chiefs of Staff meetings and acting as his go-between with the military chiefs. Churchill called Ismay his 'eminence khaki' and told him that they must be very careful not to define their powers too precisely. He did not want to create a rulebook that hemmed in his authority. He wanted to range widely and without restraint across every aspect of the conduct of the war.[7] And this he did. Ismay became a faithful supporter and confidant of Churchill and he was one of the few who came into post in May 1940 and stayed with Churchill throughout the next five long and immensely challenging years. Their relationship was good, sometimes strained, but enduring. Churchill's affectionate nickname for Ismay was 'Pug'.

The War Cabinet was at the heart of the government. Churchill reduced it from nine to only five members, although later it increased in size. As we have seen, Chamberlain became Lord President of the Council. Clement Attlee, the leader of the

Labour Party, was Lord Privy Seal. Halifax remained Foreign Secretary. And Arthur Greenwood, the deputy Labour leader, became Minister without Portfolio. The ministers responsible for each of the services were not permanent members of the War Cabinet – another mark of the centralisation of military power and decision-making to Churchill himself.

The service ministers reflected the coalition nature of the new government. Sir Anthony Eden, the Conservative ex-Foreign Secretary who had resigned in 1938, was appointed Secretary of War. Sir Archibald Sinclair, leader of the Liberals as well as Churchill's good friend and his second-in-command in the trenches in the First World War, went to the Air Ministry. And Labour's A.V. Alexander replaced Churchill himself as First Lord of the Admiralty.

In less than a week, Churchill had appointed all thirty-four members of his government. The other key appointments here included Ernest Bevin, a trade union leader and for his day a real 'man of the people', as Minister of Labour. This was a brilliantly inspired appointment, and Bevin's total commitment to the war effort and to national unity helped keep many working people on side throughout the hardships that would follow. Lord Beaverbrook, the close friend of Churchill's from his 'court' at Chartwell, was appointed to the newly created post of Minister of Aircraft Production. The Canadian newspaper magnate was not a popular figure and many people were amazed at his appointment, but Churchill realised the vital need to increase the rate at which aircraft were being built in British factories, particularly the new fighter planes like the Hurricane and especially the Spitfire, which was suffering from huge delays in production. Churchill needed a human dynamo to create a flow of electricity to speed things up, and Beaverbrook provided the necessary spark. Sir John Anderson, an experienced administrator who had organised the evacuation of children at the beginning of the war, became Home Secretary. He gave his name to the corrugated-iron bomb shelters that almost everyone who had access to a garden would construct. Herbert Morrison, who had been the Labour leader of the London County Council and was a rising star within the party, was appointed Minister of Supply. Coincidentally, he gave his name to another type of bomb shelter.

Churchill also gathered around him a group of friends and advisers whose judgement he trusted and whose company he enjoyed. They had various jobs in what now became the court of King Winston, and they came nearest to a sort of central staff. Today, they would all become paid government consultants. Then, they remained on the fringes of government. But in essence Churchill created a private office in Downing Street that reported directly to him and served his interests and curiosity. At the centre of this would be his old friend Professor Frederick Lindemann (who later became Lord Cherwell), the rather pompous Oxford scientist universally known as 'the Prof'. He would meet with Churchill almost daily and advise him on a broad range of topics over the next few years. Another member of Churchill's inner team was his son-in-law, Duncan Sandys, who took on various roving missions. Brendan Bracken, another trusted friend from the old days, acted as a wide-ranging political adviser. Desmond Morton, the man who had supplied Churchill with valuable secret information during his wilderness years, became a sort of intelligence supremo. Sitting adjacent to the Cabinet Office, he had direct access to Churchill and acted as the Prime Minister's go-between with the various intelligence agencies. He was outgoing, with a strong sense of humour, and his raucous laughter echoed down the corridors of Downing Street. John Colville later described how these advisers arrived like 'Horsemen of the Apocalypse' in the government set-up.[8] Churchill's War Lab was beginning to take shape.

The military heads of the three services, known as the Chiefs of Staff, would meet daily in the morning. They constituted a sort of battle headquarters in London to assess and advise on the strategic direction of the war. Churchill would not attend these meetings as a matter of routine, but as Minister of Defence he could chair them if circumstances required it. Ismay, a member of the Chiefs of Staff Committee, would report daily to Churchill on the meetings and the issues that were discussed. Reporting to the Chiefs of Staff Committee were the Joint Planning Staff and the Joint Intelligence Committee. The latter group assembled and assessed the best intelligence about enemy actions and intentions. This was then distributed to Churchill, the War Cabinet and the Chiefs of Staff. This system was designed so that all the major decision-makers should have possession of all the key information.

The Chief of the Imperial General Staff (CIGS), a title inherited from the heyday of Empire, was another pivotal figure in the War Lab. When Churchill became Prime Minister, General Edmund Ironside was in the post. He was an impressive, dashing figure who in his youth had provided the model for Richard Hannay in John Buchan's *The Thirty-Nine Steps*. But he had fallen out with Churchill during the Norwegian debacle because he could not bear what he saw as the constant meddling of a politician who seemed to want to run a military campaign as though he were the field commander. A fortnight after Churchill took power, Ironside was replaced as CIGS by Sir John Dill, a tough Ulsterman greatly valued by Churchill for his abilities and strategic knowledge. But Dill seemed tired and lacked the vigour that was expected of him at this point of the war. Over time, Churchill would drive Dill to distraction with his constant attention to the sort of details the CIGS considered inappropriate for a Prime Minister to concern himself with. And Churchill would grow frustrated with Dill's slowness to respond, nicknaming him 'Dilly-Dally'. But this was in the future.

Under Chamberlain, the management of war policy had been confused and poorly thought through. Ironside commented in April 1940: 'Strategy is directed by odd people who collect odd bits of information. This is discussed quite casually by everyone.'[9] This was clearly not the way to run a war. Churchill, who had not only been at the heart of the management of war policy in the First World War but who had also seen how decision-making came about (or didn't come about) in the first eight months of the Second, brought in a new broom to sweep away Chamberlain's structure and make room for his own. Placing himself firmly in the coordinating chair, he established various groups who reported in with information and advice. The War Cabinet was responsible for the ultimate direction of war policy. The Chiefs of Staff reported to Churchill daily through Ismay. A new group known as the Defence Committee (Operations), consisting of the service ministers and the service chiefs, sent top-level strategic advice up the system. Another group, the Defence Committee (Supply), concerned itself with essential aspects of running and planning the war effort in a way that left the service chiefs free to concentrate on strategic thinking.

This might all sound like administrative game-playing at best or empire-building at worst. But these new structures, creating a new 'organogram' of power, were much more than this. One participant described the practical effects as 'revolutionary'.[10] It was essential that key information should flow in the most sensible way to those who needed to know what was going on and could then make sensible decisions in the light of this knowledge. It was essential to be able to focus first on the macro (or strategic) perspective on how to conduct the war. And then to see this through by attending to the micro (or

tactical) detail of how to pursue it wisely. Finally, it was necessary to ensure that supply and logistics were focused on
achieving and supporting the strategic war aims. At the centre
of this new web of power was of course Churchill himself, in
his joint role as Prime Minister and Minster of Defence. On him
would fall the burden of understanding the big picture and
deciding on myriad details that would make the war effort
effective. Churchill now had to show how his years of military
experience as a soldier and a deep understanding of military
history would make him suitable as war leader for a nation in
crisis.

Churchill committed himself to the tasks ahead with
immense energy. An endless stream of instructions, requests,
exhortations and diktats began to flow from the top, usually in
the form of Churchill's increasingly familiar minutes. Rarely
in history have so many orders come from the desk of one
man. At no point in the Second World War was there anything
comparable – not from Stalin after the invasion of the Soviet
Union in June 1941, nor from Roosevelt after Pearl Harbor in
December of that year. Churchill's first few months in office
were more akin to President John F. Kennedy's first 'Hundred
Days', and although there were grave crises in both, the nation's
survival itself was not at stake when Kennedy took office. Over
the next few weeks and months, Churchill was to enquire about
and instruct on an enormously wide range of issues. What
progress is being made with rockets, with sensitive fuses, with
bomb sights and with Radio Detection-Finding? Can Turin
and Milan be bombed from England? Can more trees be felled
to reduce the reliance on imports and save shipping? Can
regular troops be moved from India? Can a better reserve be
built up in the Middle East? How are the coastal watches and
coastal batteries being organised? How are harbour defences
being built up? These minutes were often known by his staff

as 'Churchill's prayers', because they usually began: 'Pray tell me . . .' or 'Pray explain why we cannot . . .'. To the Chiefs of Staff he would send directives for them to consider. His officials gave these their own names and references. Hence, a long and detailed directive beginning 'Renown awaits the Commander who first . . .' became known simply as 'Renown Awaits'. It was through these directives that Churchill tried to shape the conduct and strategy of the war.[11]

As with so much of Churchill's personal style of leadership, no detail was too small to escape his attention or interest. For instance, after Eden had suggested the creation of Local Defence Volunteer forces across the country – to utilise the skills and enthusiasm of ex-servicemen who were too old or otherwise unable to join the military but who wanted to 'do their bit' – Churchill wrote: 'I don't think much of the name "Local Defence Volunteers" for your very large new force. The word "Local" is uninspiring . . . I think "Home Guard" would be better. Don't hesitate to change on account of already having made armlets, etc.' And so, with the dash of a minute, Churchill created a name that became legendary in the collective British memory of the war.[12]

Churchill was a hard taskmaster and wanted things done his way. This was his prerogative as Prime Minister. What particularly frustrated a maverick like Churchill was the slowness of officialdom. Fighting a war with the intensity he now was, Churchill wanted action and he wanted it now – better still yesterday. He developed a variety of techniques to push things through and to attribute priorities. His dispatch box always had to have documents filed in a particular way, with the most urgent military matters at the top, and documents for reading at leisure (probably over a weekend) marked with an 'R' for 'Recreation' at the bottom. Officials had to act on items he had signed or authorised. And nothing typifies his desire to get things moving more than the red sticker he had specially made marked 'Action This Day'. When he attached this to a document or a minute the instructions or requests within were to be given top priority. Churchill wanted to hear back on it within the day.

Working for Churchill must have been extremely difficult. No doubt many of his minutes sidetracked exhausted officials from other duties. But Churchill also brought an immense, restless energy to the top of government. This had been badly lacking under Chamberlain, and almost certainly would have never happened under Halifax. As all management training makes clear today, change needs to start at the top. The chief executive has to be the model and to set the pace. Churchill instigated a brand of 'change management' (to use a phrase that would have meant nothing to the old man himself) that was dramatic, heartfelt and strikingly effective. Just when Britain needed it, the pace of government went up a gear. Lord Normanbrook, a senior civil servant in the Cabinet Office, wrote later:

> This stream of messages, covering so wide a range of subjects, was like the beam of a searchlight ceaselessly swinging round and penetrating into the remote recesses of the administration so that everyone, however humble his range or his function, felt that one day the beam might rest on him and light up what he was doing. In Whitehall the effect of this influence was immediate and dramatic. The machine responded at once to his leadership. It quickened the pace and improved the tone of administration. A new sense of purpose and urgency was created as it came to be realised that a firm hand guided by a strong will was on the wheel. Morale was high.[13]

Churchill's working day reflected this full-on approach, and it soon adopted a particular and rather unusual routine. Of course, every day was different. But even when travelling and working abroad Churchill would try to keep to some of his established routine. He would be woken every morning at about 8 a.m. Breakfast and the day's papers would be brought to him in bed on a tray. Along with these came the daily report from the Map Room. This was of vital importance to Churchill, who wanted to be kept up to date on the latest movements of armies and ships around the world. The report was often delivered in person by the naval officer in charge of the Map Room, Captain Richard Pim, who had been running it since Churchill's days at the Admiralty. Then, wearing his favourite dressing gown, which was green and gold with red dragons on it, he would sit in bed, propped up by pillows, and work on his boxes. He would read messages and communications, Cabinet papers, Foreign Office telegrams, military reports and so on. Throughout the morning, a female secretary or typist would be on hand to take down his dictation straight on to a specially made silent typewriter (the sound of punching the keys annoyed him). There was rarely time to take shorthand and then type it up later. It all had to be typed and signed immediately. The typists were led by Kathleen Hill, who knew his fancies and foibles inside out. The secretaries who joined his staff during the war at first found him a terrifying figure who would call out 'Gimme my box', 'Gimme Pug', which meant call for General Ismay, or 'Gimme Prof', which meant call for Professor Lindemann. On other occasions, he would just hold out his hand and say, 'Gimme', expecting the poor woman to know exactly what was meant, sometimes to pass a black pen, sometimes a red pen, a paper punch, which he called a 'Klop', some blotting paper, or the red 'Action This Day' label to stick on the front of a minute.[14]

Churchill would stay sitting up in bed, working and dictating, for as long as possible. Most mornings, the Chiefs of Staff would meet without the Prime Minister and Ismay would arrive to report back on their deliberations. Other officials or intimates might attend him in his bedroom. Several cigars would be smoked through the course of the morning, although most went out as Churchill's thoughts or dictation distracted him, and much fuss and effort would be devoted to relighting them. Often the War Cabinet would meet in the morning, daily in the early stages of the war and twice-weekly later. Half an hour before whatever meeting required his presence, Churchill would take a hot bath and would dress with the assistance of his valet, Frank Sawyers. Throughout the war Churchill kept up the Victorian habit of relying on a valet, and Sawyers was never far from his master, in London, at Chequers or abroad. Sawyers was short, plump, bald and, in the way of old-style servant relationships, utterly devoted to his boss. Churchill would always go out in public dressed immaculately in striped trousers, a black jacket with waistcoat, a spotted bow tie and a puffed handkerchief in his top pocket. However, if there were no meetings scheduled, he would often prefer to wear his siren suit, a one-piece boiler-type suit with a zip up the front, in Air Force blue. His staff called it his romper suit.

Lunch was usually served at 1.30 p.m., and provided an opportunity for Churchill to meet and talk with military personnel, politicians or his close set of advisers and friends. The afternoon was nearly always spent in meetings of one sort or another. Churchill had a private study next to the Cabinet Room in 10 Downing Street. Meetings could be held there or in the Annexe, a specially prepared set of rooms a few minutes' walk away at Storey's Gate. The Churchills had a small flat there as well. Below the Annexe, down a spiral staircase, were the Cabinet War Rooms, an underground complex supposed to be secure from bombing. This network of offices, communications, sleeping accommodation and a Cabinet Room were designed to provide all that was needed should central government have to carry on below ground. Churchill's Map Room was also located there.[15]

Every afternoon, as soon as possible after lunch, Churchill would take a nap for about an hour. His closest officials remarked on how quickly he could fall into a sound sleep, even during the most stressful times. This mid-afternoon rest did him immense good, and after it was over he would rise, take another hot bath if there were time, and like a giant awakened would be ready to start almost another full day's work, with the typists again standing by for dictation and officials hurrying in and out as required.

Dinner was always served around 8.30 p.m. This was another opportunity to meet and talk with senior officials or leading soldiers, sailors and airmen. It was always a big occasion. Reasonable amounts of wine would be served. The meal was four courses, beginning with soup. The soup course was invariably a restorative to Churchill. Whatever frame of mind he was in, no matter how much pressure he was under, his mood would lift as he ate the soup, and dinner usually became a lively, amusing and provocative affair. Most people who describe having dinner with Churchill talk of the countless witticisms and one-liners that he came up with during the meal. Strategy would be openly discussed, new ideas and potential solutions to challenges kicked around, and politics and military affairs freely debated. From about ten o'clock onwards, Churchill would then conduct more formal meetings or conferences, often with his CIGS or the other military chiefs. After these had ended, he would routinely dictate into the early hours – more minutes or directives to the Chiefs of Staff, or outlines for an upcoming speech or broadcast. He enjoyed these late hours and clearly some of his best ideas came to him then. General Sir Alan Brooke later said that Churchill was the sort of man who had ten ideas a day, only one of which was any good, but he never knew which one.[16]

The long hours were a strain on his staff. 'It's amazing how quickly you get used to going to bed at 2.30,' Elizabeth Layton, one of the secretaries, was told before she started to work for Churchill. His top military men, none of whom had enjoyed the benefit of a mid-afternoon siesta, often found the late-night sessions exhausting, but sometimes exhilarating too. Churchill would often use the tête-à-têtes as an occasion to make up after a demanding day or to make someone feel wanted who had been bullied or cajoled by his incessant demands. No matter how grumpy or stressed Churchill had been during the day, in the small hours he would always turn to his secretaries with a glowing smile and say, 'Good night and thank you,' making them feel appreciated. The working day could end at any time from about two in the morning onwards and sometimes as late as four. This schedule was kept up seven days a week. At Chequers there was often a film show after dinner, but otherwise the routine and the workload at the weekend remained the same. It was not an easy schedule to keep up with.[17]

Despite his authoritarian control over affairs, Churchill was no dictator. He dominated conversations, and his ubiquitous role at the centre of the political and military machinery of the war meant that his influence was immense But he listened to his professional advisers. He wanted them to argue their position forcefully and he relished the cut and thrust of debate. Often strong words were used and tempers frayed, but Churchill was well used to this from Parliament and throughout his life he rarely bore a grudge. So if his military chiefs stuck to their guns and opposed him, he would always back down. Not once during the war did he overrule his military advisers on a military question.[18]

At last, Britain's war effort was losing the complacency and hesitation that had characterised the Chamberlain era. At last, the country had a leader with the knowledge, the energy and the determination to get the machinery of war moving. But events now taking place in Belgium, Holland and France would soon show how limited Churchill's impact could be in this first phase of the war.

A major priority for Churchill as he threw himself into the business of leading the nation was to follow and to give every support to Britain's ally in the war waging across the Channel. He made five visits to France over the next six weeks, and sometimes these exposed him to considerable personal risk. It was characteristic of Churchill's style of leadership that he wanted to be in the thick of it. As Prime Minister, he would travel far more extensively than any of the other war leaders. Stalin left the Soviet Union just twice – to attend the Teheran and Potsdam summits. Roosevelt travelled abroad three times during the war, but in the main he expected Churchill to come to him, which was not particularly surprising bearing in mind his own disability and the fact that he was after all a head of state while Churchill was merely head of government. Hitler hardly ever travelled far from his favourite haunts in Bavaria, East Prussia and Berlin. Churchill strained every nerve and tried every trick to support the French in their life-and-death struggle with Germany. But his effectiveness here was severely limited by factors over which he had no control.

In the spring of 1940, French military thinking was still dominated by the experience of the First World War. The French Army was haunted by the memory of one and a half million deaths. It could think only in terms of defence, hence the vast expenditure on and commitment to the Maginot Line. Most of its senior commanders had obtained all their military knowledge between 1914 and 1918. There was little or no coordination between the army and the air force. The French Commander-in-Chief, General Maurice Gamelin, had no radio communication with his senior officers from his command headquarters at the Château de Vincennes. From there he conducted affairs in splendid isolation from day-to-day events, in an echo of Great War generals commanding their armies from comfortable châteaux miles behind the line. There was no clear delineation between the roles of Gamelin and his deputy, General Alphonse Georges, who held executive command over troops along the entire front. The whole French command structure was tangled and confused. At one critical point as the Germans swept into France, Gamelin began one of his orders with the apologetic words: 'Without wishing to intervene in the conduct of the battle in progress . . .' But what else should the Commander-in-Chief be doing?

Politically, too, France was weak and divided. Paul Reynaud had been Prime Minister for only seven weeks. And he had no time for Gamelin, whom he had tried to remove from command the day before the German onslaught overwhelmed them. The French had many skilled troops and some outstanding armoured units. But, as with so much in the French Army of 1940, the tank units were slow, poorly structured and ineffectively deployed on the battlefield.

The German Army, the Wehrmacht, on the other hand, was a thoroughly modern and effective fighting force. It had been almost entirely restructured during the 1930s, learning all the lessons of defeat in 1918. German military thinking incorporated the latest techniques for using fast-moving, armoured units in sharp hammer-blows against a narrow enemy front. Although the highest level of command was full of generals who doubted that France could be knocked out in a single, rapid blow, the middle-ranking officers who led men in battle were both highly motivated and trained to seize the initiative as and when they could. For instance, General Heinz Guderian was a great enthusiast for the armoured-warfare theories proposed by British strategists like Basil Liddell Hart and tank pioneers like Generals Fuller and Hobart (theories largely ignored by the British Army of 1940). Guderian commanded a panzer corps in the assault on Belgium and made it clear to all his senior officers that their overall objective was to force their way to the Channel coast in northern France as quickly as possible. They were to use their own initiative whenever necessary so long as it contributed towards this goal.

The British Army had sent a small expeditionary force to northern France, just as it had in 1914. The troops were well trained and reasonably well equipped. But as in the First World War, they were too few in number to make a substantial difference to the battle. Their commander, Viscount Gort, was a brave, courageous general who had won a Victoria Cross and a host of other commendations during the First World War. But although he had Churchill's confidence, he could not make up for years of peacetime neglect of the army. There was also confusion over how the British troops should fit into the French operation. Gort did his best to act as a loyal ally, but he was astonished and a little ashamed at how quickly French morale collapsed.

Despite the contrasting strengths and weaknesses of the Allied and Nazi units, the outcome of the battle that unfolded in May and June 1940 was by no means a certainty. The combined troops of the French, British, Belgian and Dutch armies slightly outnumbered the Wehrmacht. And the Allies even had more tanks than the Germans. But the German Air Force, the Luftwaffe, far outnumbered the combined Allied air forces, and no French fighter could match the Messerschmitt Me-109. The RAF Hurricanes and Spitfires were formidable flying machines, but as far as the French were concerned there were never enough of them. In addition, the Germans used their

Stuka dive-bombers brilliantly in close ground support. These aircraft had a siren fitted below the wings which emitted a screaming sound as they dived to drop their thousand-pound bomb, making them daunting terror weapons. Even the most experienced soldiers broke under bombardment from a wave of Stukas.

Using a strategy of fast-moving armoured units comprising massed phalanxes of panzers, with close air support for the ground advance, and choosing to fight over a narrow rather than a broad front, the German Army now rolled out its techniques for blitzkrieg or lightning war. This combination of surprise, speed, weight and numbers had been used very effectively in Poland. In France and Belgium, the Germans were up against tougher troops in much larger numbers, but the end result was the same. The Wehrmacht had been forged into a brilliant fighting machine with an iron sense of purpose and self-confidence. The Germans totally outperformed the Allies and enjoyed one of the most spectacular victories of the war.

From the beginning of their assault at dawn on 10 May, the initiative lay with the German troops. In a brilliant and daring action, about three hundred glider troops captured the giant, modern Belgian fortress at Eban Emael, reputed to be the strongest fort in the world. The well-defended position was overwhelmed in just a few hours. Simultaneously, a complete army group moved into Holland in the north, drew out the Dutch and Belgian armies and tempted the British forward, too. But this was part of the German master-plan to draw Allied forces away from the main strike, which fell in the centre of Belgium. Here, ten panzer divisions launched a lightning attack through the Ardennes forest, which the French had believed to be impenetrable and had left largely undefended. In a series of actions from 13 to 16 May along the banks of the river Meuse, the panzers crashed through French defences and established bridgeheads on the west side of the river.

After six days, the Dutch government gave in. The royal family escaped on a British destroyer and King George welcomed Queen Wilhelmina at Liverpool Street Station. Soon the Dutch government followed the Queen into exile in London. The breakneck speed of the German advance was beyond the comprehension of French generals who still thought in First World War terms of a few hundred yards lost or gained in a day. Their poor communications and divided command hampered effective resistance. On the eve of the sixth day of fighting, with the panzer breakout along the Meuse threatening to divide his entire army, Gamelin ordered a full withdrawal of all his troops from Belgium. Prime Minister Reynaud gave the order for the government to prepare to evacuate Paris and move to Tours. He rang Churchill in an 'excited mood' and said, 'We are defeated. We have lost the battle. The way to Paris lies open.' He asked Churchill to send all the troops and planes he could.[19]

On the following afternoon, Churchill flew to Paris. At 5.30 p.m. a historic meeting took place at the Quai d'Orsay. Reynaud, his Defence Minister Daladier and Gamelin met with the British Prime Minister, who was accompanied by Generals Ismay and Dill. Gamelin outlined the collapse of the front in a short speech to all those present. The meeting fell silent. In his best French, Churchill asked about the strategic reserve: '*Ou est la masse de manoeuvre?*' General Gamelin turned to him and with a shrug and a shake of the head replied: '*Aucune . . . Aucune*' – 'There is none.' Churchill wrote later: 'I was dumbfounded. What were we to think of the great French Army and its highest chiefs?'[20]

The French pressed Churchill to send more fighter squadrons. This had already been discussed by the War Cabinet. Air

Marshal Sir Hugh Dowding was in charge of RAF Fighter
Command and had responsibility for the air defence of Britain.
He argued against sending fighters to France because this would
weaken the British air defensive system. Churchill felt that
although it was a grave risk, something had to be done to 'bolster up the French'. He telegrammed London to say: 'It would
not be good historically if their requests were denied and their
ruin resulted.' The War Cabinet met at eleven o'clock that night
to discuss this telegram and agreed to dispatch a total of ten
fighter squadrons. Churchill immediately went to Reynaud's flat
to tell him the news. After an interval, the French Prime Minister
appeared from his bedroom in his dressing gown. Daladier was
called to join them and he shook Churchill's hand in gratitude.
Churchill got to bed at the British Embassy at about 2 a.m.

Over the next few days, the German panzers broke through
and began their drive for the coast. The Allied position looked
as though it were about to crumble. Gamelin was replaced by
the sprightly seventy-two-year-old General Maxime Weygand,
who tried to rally the troops but could do nothing to reverse
the disasters that had already befallen the dispirited French
Army. On 22 May, Churchill flew back to France to meet
the French commanders at Vincennes. He was pleased to see
that Weygand was reviving the French command, but by now
the first German panzers had reached the Channel coast near
Abbeville. The Allied armies were cut in two. The advanced
panzer divisions moved north along the coast and after some
bitter fighting captured first Boulogne, then Calais.

Ironside, the CIGS, was sent by the War Cabinet to visit Gort
in the field and tell him to move his troops south-west. Gort
carried out a small counter-attack against the advancing panzers near Arras which was successful. This forced even the elite
panzers of the 7th Division, led by Erwin Rommel, to pause
briefly and regroup. But the planned French support did not

appear; and, fearing that his men would be cut off, Gort called off the counter-attack and, in defiance of his orders, began to retreat towards the coast.

At this point, General von Rundstedt, commander of the German army group in the centre of the battle, ordered his panzer units to pause. He was stunned by the speed of their advance, well over two hundred miles in little more than three days, and worried that they were now too exposed. In his diary, he noted that the enemy had been fighting 'with extraordinary tenacity', and he wanted to regroup and wait for the infantry to catch up. He was fearful of the very type of counter-attack that Weygand was in fact now planning. The German High Command debated whether to turn south and head for Paris or turn north to mop up the British and French armies first. At this point, Hitler arrived in person at von Rundstedt's headquarters. For the first time since the Battle of France had started he intervened directly and endorsed von Rundstedt's order. Maybe he thought it was more likely that the British would seek peace with Germany if their army had not been destroyed in the field. He almost certainly wanted to show his commanders that he was in control.[7] The panzers came to a halt, a few miles from the French port of Dunkirk.

On 27 May, an evacuation of British and French soldiers began around Dunkirk. The head of the Luftwaffe, Hermann Goering, told Hitler that his planes would finish off the Allied armies. And they nearly did so. Soldiers under constant harassment from the air cursed the RAF for not coming to their aid. But this was unfair. The RAF was attacking the German bomber formations as they assembled many miles away from the evacuation beaches. In a foretaste of what was to come later in the summer, the RAF pilots shot down large numbers of Goering's bombers but suffered substantial losses themselves. After a few days, the weather turned for the worse, the Luftwaffe could not fly, and the evacuation went up a gear. Boats of every shape and size came the twenty-odd miles from the Kent coast to assist the navy in the evacuation. Hundreds of these 'small ships' helped to get men off the beaches in what was now called the 'miracle of Dunkirk'. For several days and nights, the evacuation continued, with exhausted men clambering into fighting ships, trawlers, drifters, fishing boats, tugs, yachts and motor launches. The Admiralty initially estimated that they would be able to evacuate about 45,000 men. By 4 June, when the evacuation was finally called off, 338,226 men had been rescued, two-thirds of them British and one-third French, Belgian and Dutch.

Gort, the commander of British forces in France and Belgium and hero of the First World War, chose to stay with his men in the ever-shrinking defensive perimeter around the Dunkirk beaches, to face death or capture. However, Churchill thought his capture would provide the Germans with too great a trophy, so he ordered Gort to hand over to a deputy and evacuate. This he reluctantly did on 1 June, feeling bad about his escape. He believed that the criticism he later received for deserting his post was not fair as he had only been following Churchill's orders. But this was not the end of Gort's career and he would go on, two years later, to be Governor of Malta at a critical point in the war, when the survival of that beleaguered island was of vital importance.

In the last days of May, Churchill had faced another crisis. He had been Prime Minister for only two weeks, and his hold on the reins of power was by no means secure. Hitler thought Churchill would not last long and that the British would soon realise the folly of continuing to fight. Indeed, many in the British establishment were beginning to panic in the face of the disaster unfolding on the continent and the prospect of having to fight on alone. One Tory MP wrote in his diary: '"All is lost" sort of attitude in evidence in many quarters.'[22] And Halifax, very much a member of the establishment, believed that Britain should at least explore the possibility of negotiation with Hitler before facing the very real likelihood of total defeat. He made an approach to Mussolini, Hitler's ally, although at this point Italy was still not at war with Britain or France. In the War Cabinet, Halifax argued that Britain's long-term interests might be better secured by negotiating now before France fell, rather than later, under duress. The War Cabinet was divided. The two Labour members, Attlee and Greenwood, were opposed to any sort of negotiation with Hitler because they thought it would undermine the war effort if news of it ever got out. Chamberlain seems to have sat on the fence. Churchill, unsurprisingly, was instinctively opposed to any sort of parley with Hitler. 'Nations which went down fighting rose again, but those who surrender tamely were finished,' he argued.[23]

Matters came to a head on the afternoon of 28 May, when the War Cabinet discussed whether to open negotiations. Halifax was now in effect making a direct challenge to Churchill's leadership, not because he wanted to take over but because he thought the Prime Minister's thinking was wrong and a new course was needed. After an hour of passionate discussion, the War Cabinet was adjourned. Churchill had called a meeting of the full Cabinet at 5 p.m. In this he gave the twenty-five members of the Cabinet a frank account of the perilous situation in France. Hugh Dalton, the newly appointed Minister of Economic Warfare, wrote that Churchill 'was quite magnificent. The man, and the only man we have, for this hour.' He then recorded that Churchill said: 'I have thought carefully in these last days whether it was part of my duty to consider entering negotiations with That Man.' Churchill argued that Britain would not get any better terms by making peace with Hitler now than if the country went on and fought it out. Churchill continued: 'The Germans would demand our fleet – that would be called "disarmament" – our naval bases and much else. We should become a slave state.' He ended his speech by saying, 'We shall go on and fight it out, here or elsewhere, and if at last the long story [of British history] is to end, it were better it should end, not through surrender, but only when each one of us lies choking in his own blood upon the ground.' At this, a cheer went up around the table. Members of the Cabinet went over and congratulated Churchill, shook his hand and patted him on the back. Churchill was delighted by this spontaneous display of support from such a group of men, many of whom had long been his political adversaries. He later wrote: 'There is no doubt that had I at this juncture faltered at all in leading the nation, I should have been hurled out of office.' When the War Cabinet resumed its meeting at seven o'clock that evening, Churchill reported back on the strong support he had received. His resolve had been both personally and politically strengthened. There was to be no more talk of a parley with Hitler. Churchill had faced down Halifax. He had won. One historian has described this as a day that changed the course of world history, which would have been so different if Churchill had agreed to negotiate a deal with Hitler: 'Then and there he [Churchill] saved Britain, and Europe, and western civilisation.'[24]

On 31 May, in the midst of the Dunkirk evacuation, Churchill flew once again to Paris. He met with Reynaud and, for the first time since 1918, with Marshal Pétain, the legendary Great War hero who had been recalled as Deputy Prime Minister. Churchill repeated his line that Britain would fight on. In a conversation after the main meeting, he began to suspect that Pétain had now become a defeatist, and that he would agree to a separate peace treaty with Germany.

Churchill had not spent much time in the House of Commons since becoming Prime Minister. On 4 June, he reported to the

House what he described as the 'miracle of deliverance' at Dunkirk. But while celebrating the triumph of withdrawing so many men from the jaws of certain defeat and capture, he admitted: 'Wars are not won by evacuations.' Churchill hinted for the first time publicly in the speech that if France collapsed then Britain would fight on alone. In a passage that was partly intended to rally the British people to the struggle ahead, and partly intended to express Britain's resolve to overseas friends, especially in America, he concluded with some stirring phrases that have gone down in popular memory. He proclaimed:

Even though large tracts of Europe and many old and famous States have fallen or may fall into the grip of the Gestapo and all the odious apparatus of Nazi rule, we shall not flag or fail. We shall go on to the end. We shall fight in France, we shall fight in the seas and oceans, we shall fight with growing confidence and growing strength in the air; we shall defend our island whatever the cost may be. We shall fight on the beaches, we shall fight on the landing-grounds, we shall fight in the fields and in the streets, we shall fight in the hills; we shall never surrender.[25]

The speech was immensely well received by a packed House of Commons. One MP said it was worth 'a thousand guns'. In those days, Parliament was not broadcast, but the speech was very widely reported and had a great impact. Vita Sackville-West wrote to her husband, Harold Nicolson, after listening to a report of the speech on the radio: 'Even repeated by the announcer it sent shivers (not of fear) down my spine.'[26]

Belgium as well as Holland had now surrendered. The armies in the north had been defeated. Now Hitler's forces could turn south and move on Paris. On 5 June, the final phase

of the Battle of France began. The British Army had evacuated
from Dunkirk without most of its heavy weaponry, having
abandoned more than 1400 artillery pieces, 38,000 vehicles
and over 7000 tons of ammunition. There was precious little
left to protect Britain should an invasion come. But Churchill
knew that the military defeat of France would bring a political collapse. And the loss of an ally of this importance could
be a fatal blow to Britain's own war effort. So he remained
committed to the rapidly disintegrating French Army and
sent two new and well-equipped divisions to north-west
France.

Churchill requested another meeting with Reynaud to find
out what the French were thinking. By now Paris had been
abandoned and it was not easy to find a place to meet. On 11
June, Churchill and his small delegation flew to Briare on the
Loire River for a meeting of the Supreme War Council. On
arrival at the small landing strip, it was immediately appar-
ent to Churchill how bad things had got. He tried to project a
confident air. But the French were not responsive. A conference
followed that evening in a nearby château that had been
temporarily commandeered. It was woefully inadequate for
this function and had only one telephone. General Weygand
asked that every British fighter squadron should be sent
to France. 'Here', he said, 'is the decisive point. Now is the
decisive moment. It is therefore wrong to keep *any* squadrons
back in England.' Churchill erupted. 'This is not the decisive
point and this is not the decisive moment,' he declared forcefully. 'That moment will come when Hitler hurls his Luftwaffe
against Great Britain. If we can keep command of the air, and
if we can keep the seas open, as we certainly shall keep them
open, we will win it all back for you.' In this atmosphere, no
progress was made. Churchill heard after the meeting exactly
how appalling the situation had become, with troops unable

to move along roads packed with refugees and everyone under constant machine-gun fire from German aircraft. He came away from the meeting convinced that a complete French military collapse was imminent. And he feared that Pétain would sign a peace treaty with Hitler. Only a meeting with a young French colonel, Charles de Gaulle, who had been appointed Under-Secretary for National Defence, cheered Churchill. De Gaulle favoured fighting a guerrilla war, and Reynaud seemed to like this approach. Churchill thought that maybe de Gaulle would take over from Reynaud if the latter stood down. But overall he left in a state of despair.

On the flight home, it was too cloudy for the Hurricane escort to accompany Churchill's plane. He chose to fly back anyway without it as there was much to do back in London. Over the French coast, the clouds cleared and Churchill looked down on Le Havre burning, the battle raging below. At this point, his aircraft dived steeply from eight thousand to about a hundred feet. German fighters had been spotted. The manoeuvre worked, and Churchill's plane flew on back to Britain, unnoticed and unharried.[27]

Two days later, the German Army entered Paris. Hitler's divisions then turned south and west, and within days had finished off the last French resistance. The military battle was over, but the political argument continued. Churchill made one final visit to France on 13 June. This time he flew to Tours to meet the French leaders. Such was the chaos on the ground that no one came out to meet the British Prime Minister. Churchill's delegation found a car and drove into the town along streets crowded with refugees and cars piled high with family possessions. Churchill couldn't even find somewhere for lunch. Eventually a café owner was persuaded to open up and serve the British leader a light meal. When Churchill finally met

Reynaud and the other French leaders it was obvious that final capitulation was near.

Over the following days, Churchill made it clear to the French leaders that he would not release them from the agreement made earlier in the year that neither country would seek a separate peace treaty. An extraordinary plan of a union between the two countries was suggested, with joint citizenship, so France could fight on. But nothing came of this. Any thought of rescuing France had now gone. Reynaud resigned on the 16th and was replaced by Pétain. Churchill hoped that the French government would move to North Africa and continue to fight from there, taking their fleet with them. He was particularly exercised by the fate of the French fleet, which he didn't want to fall into Nazi hands as this would give the German Navy a decisive advantage in the Mediterranean. He had obtained a promise from Admiral Darlan, the French naval commander, that he would never hand his fleet over to the Germans.

On taking power on 17 June, Pétain immediately opened negotiations with the Germans for an armistice. This was effectively the surrender of France. After a few days of delays, Hitler made it clear that the French must accept the terms on the table. On 22 June, the French generals signed the surrender at Compiègne. To rub in the humiliation, the Germans made the French sign the document in the same railway carriage in which the Germans had signed the Armistice in November 1918. Events had come full circle. The northern half of France would be occupied by German troops. The southern half would be run by Pétain from Vichy as a semi-independent but pro-German state.

A few days after the fall of Paris, Hitler – accompanied by Albert Speer, his favourite architect, a newsreel cameraman, and his bodyguard of SS troopers – was driven around the sights of the French capital. As he posed for the cameras in front of the Eiffel Tower, Hitler must have reflected on what a stunning victory his soldiers had won for him. For four whole years, from 1914 to 1918, the German Army had advanced only a few miles. Now, in just twelve weeks, since early April, his forces had conquered Denmark, Norway, Holland, Belgium, Luxembourg and France. In the previous ten months he had routed the Polish, Dutch, Belgian, French and British armies. Where would his triumphant war machine take him next?

Meanwhile, Winston Churchill, barely six weeks into his premiership, had to pick up the pieces after the utter catastrophe, for that's what it was, of France's surrender. But that had not been *his* battle. He had been able only to encourage and cajole from the sidelines. In the drama of the next conflict, the Battle of Britain, Churchill and his War Lab would play a leading role.

4

Spitfire Summer

Tuesday 18 June 1940 was the 125th anniversary of the Battle of Waterloo. At the time of the Napoleonic Wars, Britain had faced the possibility of an invasion from across the Channel. Now, with France defeated, the nation faced the possibility again. The guiding hand of history was never far from Churchill's shoulder. And on this day he felt it keenly. That afternoon, he addressed the House of Commons, reflecting on France's stunning defeat and looking ahead to Britain's prospects in fighting on alone. He said: 'What General Weygand called the Battle of France is over. I expect that the Battle of Britain is about to begin.' He wanted people at home and abroad to know that Britain would and could fight the battle on land, at sea and in the air. 'I look forward confidently to the exploits of our fighter pilots,' he told Parliament, 'who will have the glory of saving their native, their island home, and all they love, from the most deadly of all attacks.' He spoke of how all four dominion governments – Australia, New Zealand, Canada and South Africa – had guaranteed their support to Britain. But there was no doubt that with such a powerful foe only twenty-one miles across the Channel from southern England, the 'dread balance sheet' predicted a hard-fought battle ahead. Churchill argued: 'I see great reason for intense vigilance and exertion, but none whatsoever for panic or despair.'

He ended his address with a Shakespearean flourish, recalling Henry V's famous speech to his troops on the eve of the Battle of Agincourt. As ever, he spoke passionately of his belief in the special role carved out for Britain and its people. Churchill said of the upcoming Battle of Britain:

Upon this battle depends the survival of Christian civilisation. Upon it depends our own British life and the long continuity of our institutions and our Empire. The whole fury and might of the enemy must very soon be turned on us. Hitler knows that he will have to break us in this island or lose the war. If we can stand up to him, all Europe may be free, and the life of the world may move forward into broad sunlit uplands; but if we fail, then the whole world, including the United States, and all that we have known and cared for, will sink into the abyss of a new dark age made more sinister, and perhaps more protracted, by the lights of a perverted science. Let us therefore brace ourselves to our duty and so bear ourselves that if the British Empire and its Commonwealth lasts for a thousand years, men will still say 'This was their finest hour.'[1]

Churchill was persuaded to broadcast this same speech to the nation on the BBC at nine o'clock that evening. This was one of his wartime speeches that would be heard not just in the House of Commons but by millions of people. The words of his flourishing finale are among the most famous Churchill ever uttered, and are almost as well known today as they were in the summer of 1940, having been repeated in countless television documentaries since. One vital part of Churchill's task over the next few months would be to rally public morale. Not only had he to lead Britain's war policy, but he must also call upon all the power of words and phrases that he could muster to encourage the nation to keep up the fight, despite what many saw as overwhelming odds against Britain's survival. The head of government had to display leadership, guidance, encouragement and the passionate inspiration to persuade the British people to fight on. This, we know, he did magnificently. Churchill really was walking with destiny now.

Ironically, to many who heard Churchill speak these famous words both in Parliament in the afternoon and then again on the radio in the evening, his radio performance came across as lacklustre. John Colville, his private secretary, thought he sounded tired and that he read the speech as though with a cigar in his mouth.[2] Harold Nicolson, Under-Secretary of State at the Ministry of Information, wrote to his wife:

> How I wish Winston would not talk on the wireless unless he is feeling in good form. He hates the microphone and when we bullied him into speaking last night, he just sulked and read his House of Commons speech over again. Now, as delivered in the House of Commons, that speech was magnificent, especially the concluding sentences. But it sounded ghastly on the wireless. All the great vigour he put into it seemed to evaporate.[3]

Of course, Churchill's style of oratory long predated the advent of broadcasting. He had grown to political maturity accustomed to large-scale public speaking and to debating within the chamber of the House of Commons. He never seemed comfortable on the radio. This was probably because he was used to addressing an audience and relied upon the vibrancy of a 'live' public address. Speaking alone in a recording studio into a small microphone must have seemed a very artificial experience to him. From the time he became Prime Minister to the end of 1940, he made only seven speeches on the radio, and some of these were quite short. Many of his most famous speeches were made only in the House of Commons. They are familiar to so many people today because he recorded them on to gramophone discs for Decca Records after the war.[4]

It is interesting to speculate whether Churchill would ever have become a persuasive television performer. Although he lived on well into the TV era, he did not use the medium much. Some rushes of different takes have survived of him delivering a party political broadcast in 1950. He is not impressive although he seems more comfortable than he did on the radio – possibly because with a film crew present at least he had an audience of sorts.

Churchill had framed the fight to come in epic terms. And it proved to be one of the decisive battles of the Second World War. There was little Churchill could do to lead the Battle of Britain itself. That really had to be left to the RAF. The government's role was to ensure that, after years of neglect, the RAF now at this eleventh hour had enough fighting aircraft and had the correct command and control systems in place to take on the Luftwaffe. But Churchill himself had two additional tasks during the Battle of Britain. First, he had to secure and stir the determination of the British people to fight on. If the sort of defeatism that had spread among the upper classes in May 1940 before the rescue from Dunkirk had gained a widespread hold upon the nation, the political will to continue the battle might be undermined. That is why his speeches over the next few months would be so vital to the war effort. Second, he had to maximise the backing Britain received from its allies. More than anything else, this meant ensuring the support of the United States.

Churchill always favoured face-to-face discussions with friends or opponents. His five visits to France during the battle there were vital for his understanding of the French position and his reading of the differing views of the French leadership. It is no surprise then that he dedicated a large proportion of his time during the Second World War to fostering and building a direct and personal relationship with the US President, Franklin D. Roosevelt. The two men had met briefly at a dinner towards the end of the First World War, when Churchill had been Minister of Munitions and Roosevelt was Assistant Secretary of the Navy. At the time, Roosevelt was unimpressed and described Churchill as a 'stinker . . . lording it all over us'.[5] To his embarrassment, Churchill later forgot this early meeting. However, he sent Roosevelt signed copies of his Duke of Marlborough books, and Roosevelt's son visited him at Chartwell in 1933. When Churchill became First Lord of the Admiralty in September 1939, Roosevelt opened what would become a remarkable, historic correspondence. Roosevelt wrote to congratulate Churchill, saying: 'I shall at all times welcome it if you will keep me in touch personally with anything you want me to know about.'[6] This direct approach from the US President to a single government minister was highly unusual, to say the least. Roosevelt thought Churchill would soon become Britain's wartime leader and he wanted to get his hand in early.[7] Churchill asked the Prime Minister for permission to begin his own correspondence via telegram with the President. Chamberlain, who was keen to use any route to transmit the government's views to the White House, agreed. So Churchill began a long correspondence with Roosevelt characterised by his jokey signature on every telegram: initially 'Naval Person' and from the time he became Prime

Minister 'Former Naval Person', signifying the bond between them from the days when they were responsible for their navies.

Throughout the winter and early spring of 1939–40, Churchill telegrammed Roosevelt every few weeks to keep him up to date with British government thinking. Every message had Chamberlain's approval. In February, Roosevelt replied to one of Churchill's telegrams with the words: 'I wish much that I could talk things over with you in person – but I am grateful to you for keeping me in touch, as you do.'[8] He obviously appreciated this line of communication from a man he respected at the centre of the British government. And Churchill was well aware of their value, too. The US Ambassador in London at the time was Joseph Kennedy, the father of John, Robert and Edward. From a staunchly Irish-American clan, Kennedy was an Anglophobe who despised the British Empire and thought that British society was hopelessly antiquated. During the critical moments in 1940, he was convinced that Britain would not pull through, that its military would be defeated, and that Churchill, whom Kennedy thought epitomised everything that was old-fashioned and bad about Britain, would seek out a peace deal with Hitler. So Churchill's direct line to Roosevelt was a way to get a very different message directly to the heart of Washington.

When Churchill became Prime Minister in May, the tone of the correspondence changed. It became more personal and over time more friendly. But Churchill still wanted to get tough messages through to the White House. He wanted Roosevelt to realise the dire consequences of the current European events for US interests. On 15 May, Churchill sent his first message to Roosevelt as Prime Minister in continuation of what he described as 'our intimate, private correspondence'. He did not mince his words, telling the President:

If necessary, we shall continue the war alone, and we are not afraid of that. [This was in fact the first time the thought of continuing the war if France were defeated was raised.] But I trust you realise, Mr President, that the voice and force of the United States may count for nothing if they are withheld too long. You may have a completely subjugated Nazified Europe established with astonishing swiftness, and the weight may be more than we can bear.

The telegram went on to ask for the loan of forty or fifty destroyers, several hundred aircraft, anti-aircraft guns and ammunition, and a supply of steel.[9]

The American response was swift. Hundreds of aircraft, half a million rifles, and nearly one thousand 75mm guns were shipped up and sent across the Atlantic in less than a month.[10] The loan of destroyers was a more complex issue, and Roosevelt thought this would need approval from Congress, which he was unlikely to get. Churchill insisted in his next message: 'if American assistance is to play any part it must be available soon'.[11] Throughout the summer of 1940, Roosevelt did what he could to support the fragile British position. But he was constrained by several factors. As a neutral, under US law, he could go only so far in supplying arms to a belligerent nation without requesting the agreement of Congress. Here various US domestic factions came into play to balance out the widespread support felt for Britain. The German-American, Italian-American and even Irish-American groups were all opposed to offering aid to Britain. And Roosevelt faced a presidential election that autumn. To boost his chances he made several declarations guaranteeing that the United States would not get involved in the war raging in Europe. Churchill kept up the pressure on Roosevelt and spoke of 'the Common Cause' shared by the two English-speaking nations. He kept going back to request the urgent loan of destroyers, which he wanted to bolster Britain's defences against the submarine menace and against a possible invasion by sea. His messages included phrases like 'The need is extreme' and 'in the long history of the world this is a thing to do *now*'.[12] He was infuriated by the inability of many senior US officials to see the importance of this. Eventually, in August, Churchill's pleading paid off and a deal was struck whereby the United States agreed to loan Britain fifty antiquated destroyers in return for a lease on eight British naval bases in the Caribbean and the Atlantic. The Americans also insisted that if Britain were defeated, it would not surrender its fleet to the Germans but would send it abroad to defend other parts of the Empire. This deal, which later formed the basis of the Lend-Lease Agreement, clearly marked a move by the US towards an alliance with Britain, and it was a sign that the US would provide extensive military aid – almost everything short of joining the war.

The critical point is that through this correspondence, a direct line between Downing Street and the Oval Office, Churchill knew that he had the sympathy of the President and military support from America as far as it could go within the terms of US law. This 'special relationship' would grow and blossom as the war developed. There would be nearly two thousand telegrams, several hundred phone calls, and many person-to-person meetings and summits. But in the summer of 1940, that lay in the future. For now, Churchill's confidence, thanks to his secret correspondence, that he enjoyed Roosevelt's support gave him the courage and resolution to carry on the fight against Hitler.

One of the things that most agitated Churchill after France's defeat was the position of the powerful French fleet. As we have seen, Churchill had obtained a promise from Admiral Darlan the French naval commander that he would not hand the fleet over to the Germans. But Hitler had insisted on the

surrender of the fleet in the terms of the armistice, and Pétain
and Darlan, now Minister for the Marine in the new Vichy government, were determined to keep to the terms they had signed
with the Nazis. The French Navy was scattered, and some of it
was already in British hands. But a substantial battle fleet was
now in Algeria, at the port of Mers el-Kebir. If this were used
against Britain, it might turn the naval war almost overnight.
The British Chiefs of Staff reviewed their options and offered
the local French naval commander various choices: he could
hand his warships over to the Royal Navy; sail to the West
Indies and remove them from the war; or scuttle them. From
Vichy, Darlan ordered his fleet commanders not to breech the
terms of the armistice and to have no truck with the British
demands. After several days of mounting crisis, an order was
sent from London to Admiral Sir James Somerville, in command of Force H off Mers el-Kebir.

At 5.54 p.m. on 3 July, the heavy guns of the British
battleships of Force H opened fire on the French warships in
their Algerian harbour. The bombardment lasted for only about
ten minutes. In that time the Royal Navy sank two French
battleships and ran aground one of the newest French battle-
cruisers. Elsewhere, British naval aircraft attacked French
battleships, putting them out of action. In total, twelve hundred
French sailors were killed. Churchill called it 'a hateful decision'
to open fire on men who had been allies only weeks before, 'the
most unnatural and painful in which I have ever been concerned'.[13] But no one in the War Cabinet or among the Chiefs of
Staff had hesitated when it came to sending the order to open
fire. The French fleet simply had to be put out of action to prevent the Nazis from using it against Britain. Around the world,
and especially in Washington, governments saw the decision to
destroy the fleet as a sign of the ruthlessness of the British war
leadership to fight on with vigour and determination. It was a

painful episode, but Churchill was convinced that it improved Britain's standing at a time of deep national crisis.

On the same day, the Soviet Ambassador to London, Ivan Maisky, paid Churchill a visit at Downing Street. Maisky describes Churchill as being 'full of life and energy . . . cheerful and fresh'. He asked the Prime Minister what was his general strategy, now that France had fallen. Churchill drew on his cigar and replied with a smile: 'My general strategy at present is to last out the next three months.'[14] Churchill knew that a German invasion could be attempted only during the summer months. Once autumn and winter set in, the opportunity would have passed. It was a modest target, to survive the next three months, but it was a realistic one. And it would not be easy.

Hitler's success in Northern Europe had not only astonished his enemies. It had surprised even his own generals. They were now left looking across the twenty-one miles of sea that separated northern France from the cliffs of Dover, wondering what to do next. There was no military plan in place for an invasion of Britain. But on 16 July, Hitler issued Führer Directive No. 16. In this he said: 'As England, in spite of her hopeless military situation, still shows no sign of willingness to come to terms, I have decided to prepare and if necessary to carry out, a landing operation against her.' He ordered his Armed Forces Supreme Headquarters to prepare an invasion plan. General Franz Halder, Chief of Staff, was put in charge. He saw Operation Sealion, as the projected invasion was called, as little more than a major river crossing on a broad front. Grand Admiral Erich Raeder had a different view. After the Norway campaign, he knew he could not match the firepower of the Royal Navy. He feared British warships would wreak havoc if they got in among his invasion fleet. Precious weeks were lost while the German High Command debated their options.

It was Hermann Goering – First World War fighter ace, Reichsmarschall, and probably the second most famous figure in the Nazi Party – who persuaded Hitler which course to follow. He said simply leave it to the Luftwaffe. His planes had already started a war of attrition against the RAF. He predicted that in four weeks he could force the British to surrender. Goering's argument helped to resolve the concerns of the German Navy. By destroying the RAF and winning air supremacy for Germany, the Luftwaffe would bring Britain to its knees. Maybe then an invasion would not even be necessary.

So, despite planning for an invasion, Hitler was confident he could force the British government to negotiate a peace deal. On 19 July, in a speech in Berlin, Hitler offered the British people a choice between peace with Germany or 'unending suffering and misery'. The mood in Britain was now far more defiant than it had been just two months before. The War Cabinet did not even discuss the matter. When Churchill was asked if he wanted to make a response, he wrote: 'I do not propose to say anything in reply to Herr Hitler's speech, not being on speaking terms with him.'[15]

The later myth created around the Battle of Britain was that the nation stood alone, hugely outnumbered by a superior and mighty foe. This ignores the fact that Britain was then at the head of a vast empire. Much of the world map was still coloured red. And pilots from around the Empire, especially from Canada, South Africa and New Zealand, rallied to the cause and fought bravely throughout the summer of 1940. In numerical terms, the RAF was outnumbered by roughly two-to-one: the Germans could call on about 1400 bombers, 300 dive-bombers, 800 single-engined Messerschmitt Me-109 fighters and about 240 twin-engined Messerschmitt Me-110 fighters, whereas RAF Fighter Command had only about 500 serviceable Hurricanes and Spitfires. But Lord Beaverbrook, as Minister of

Aircraft Production, was already making real increases in output by introducing round-the-clock shifts. And the factories in the Midlands and the South were soon producing about four hundred aircraft every month, enough not only to replace the losses but to grow the overall number of planes available. The problem would be the supply of trained pilots.

RAF Fighter Command had one vital advantage in the summer of 1940. Since Churchill had joined the secret Air Defence Research Committee in 1935, research into Radio Detection-Finding (later known by the American term 'radar') had developed considerably. During 1938 and 1939, British scientists based at Bawdsey on the Suffolk coast had designed and constructed a line of radar stations along the southern and eastern coasts of Britain called the 'Chain Home system'. This was an early warning system that enabled the RAF to spot and plot the course of German bombers and their fighter escorts as they crossed the Channel. Each radar station consisted of a set of giant towers that sent radio pulses out over the sea. At the base of the towers was a receiver hut where skilled operators, often members of the Women's Auxiliary Air Force (WAAF), would stare at cathode-ray tubes for hours on end. They measured the time it took for the radio waves to bounce back from approaching aircraft. By assessing the line of the approaching aircraft from two or more radar stations it was possible to predict with some accuracy the route of the raiders.

In the dogfights that followed, height proved a critical factor. If the RAF fighters could be at a higher altitude than the Luftwaffe bombers when they arrived over their targets, they would have the dual advantages of speed and surprise as they swept down from above, and so were more likely to inflict serious damage on the German aircraft. It took a fighter plane between thirteen and fifteen minutes from the order to scramble to get to its optimum height. But a German bomber could cross the English Channel in about five minutes. The key advantage provided by radar was that it gave the RAF the extra minutes it needed to get its fighters into the air and above the German bombers with accurate information as to where they were heading.

But radar was not the only vital ingredient in the RAF early warning system. The technology was still very basic. It worked well over the sea but not over the contours of land. So a second line of defence was formed by building a network of more than a thousand observation posts. From these, observers armed with little more than binoculars and aircraft recognition books identified and reported movements of enemy aircraft and telephoned the information to control rooms. At the peak of the Battle of Britain, more than fifty thousand members of the Observer Corps were in action reporting vital details of German aircraft movements.

The information provided by radar stations and the Observer Corps around the country was only as good as the intelligence interpreted from it. Air Marshal Sir Hugh Dowding had been Chief of RAF Fighter Command since July 1936. His encouragement of the development of radar and his conviction that all-metal, single-wing aircraft were the way of the future helped to shape RAF fighter defences by the time of war. But, with the input from some prominent scientists, Dowding also created a command and control system in the late 1930s that was able to analyse, interpret and focus the information from radar in a brilliantly effective way. The thousands of separate reports that came in from the radar stations and ground observers were fed to the Operations Room at RAF Fighter Command headquarters at Bentley Priory, near Stanmore in north London. Here the information was processed in filter rooms and then plotted on to a huge table map of southern and eastern England.

The Young Winston

Churchill in his smart uniform in the 4th Hussars. But the cost of uniforms, horses, mess bills far exceeded the income.
(Getty Images)

Churchill in 1904, the year he left the Conservatives and joined the Liberals. As a young politician, Churchill was a member of one of the great reforming governments of the twentieth century. (© Hulton-Deutsch Collection/Corbis)

Churchill as prisoner of war, Pretoria, November 1899. He later wrote that he hated his period in captivity more than any other period in his whole life. (TopFoto)

Churchill and Lord Fisher in 1913. Churchill brought the ebullient Fisher back to the Admiralty in 1914 with disastrous consequences.
(PA/PA Archive/Press Association Images)

Churchill checking proofs in his study at Chartwell, February 1939. His years out of office in the 1930s gave him plenty of time to develop his views on history and the 'special destiny' of the British people. (Getty Images)

Churchill leaves Downing Street after a crisis War Cabinet on the day Hitler invaded Belgium and Holland, 10 May 1940. Later that same day he became Prime Minister. (Getty Images)

RAF Fighter Command Control Room, Bentley Priory. The command and control structure created by the RAF with input from scientists like Sir Henry Tizard on the eve of war worked brilliantly during the Battle of Britain. (Imperial War Museum, IWM)

Churchill and Captain Richard Pim in the Map Room at the Cabinet War Rooms. Churchill eagerly followed troop and naval movements every day and Pim travelled with him setting up a temporary Map Room wherever he was. (IWM)

Churchill deep in conversation with Alex Henshaw, Spitfire test pilot. No doubt they were discussing the strengths and weaknesses of the great British fighter aircraft. (IWM)

Churchill visits Bristol, 12 April 1941, the day after a heavy bombing raid. His visits to bomb-damaged cities were enormously popular. (IWM)

Churchill with his scientific guru Prof Lindemann (left with bowler hat), Air Chief Marshal Portal, Admiral Pound and General Ismay, all key players in his War Lab. (IWM)

Two scientists who contributed to new ideas during the war.

Left, Sir Solly Zuckerman who studied the impact of bombing and whose work was distorted by others. (Hamish Hamilton/Zuckerman family)

Right, Alan Turing, the genius behind the 'Bombes' built at Bletchley Park to decode German Enigma signals and one of the founders of the post-war computer industry (Getty Images)

Churchill poses with a Tommy gun sent in an arms shipment from America. He loved to get his hands on new weapons. (IWM)

Highly trained young men and women wearing headsets would then move coloured markers around the table to represent enemy raiders and RAF interceptors.

From Bentley Priory, detailed information about enemy air raids was passed on to the four fighter Groups that each controlled a sector of the UK's fighter defence. Each Group had its own operational command centre mirroring the one at Bentley Priory. The Group Operation Rooms held lists of fighter squadrons and their states of readiness, from two minutes to twenty minutes, and a list of units currently in the air and in action. The decision to scramble the fighters would come from Group headquarters, as would requests for reserves during heavy raids. The fighters most often in the front line were those of 11 Group, led by a tough and determined New Zealander, Air Vice-Marshal Keith Park. His Group covered the whole of south-eastern England from its headquarters at Uxbridge in west London. This was the centre of the complex mass of communication links utilising hundreds of miles of telephone cables. The structure created under Dowding's supervision thankfully worked well during the Battle of Britain. The fate of the nation would hang on the split-second decisions made in these RAF command centres.

When all the information had been processed and the commands had been issued, it was, of course, down to the pilots and their aircraft to fight it out. RAF Fighter Command could pit two brilliant single-engined fighter aircraft against the Luftwaffe. Much has been written (and exaggerated) about the superiority of the Hawker Hurricane and the legendary Supermarine Spitfire over their German rivals. The Messerschmitt Me-109 was fast and agile, but its wings were weak and the pilots feared they would be pulled off in a fast dive or turn. This, naturally, slowed them down in combat from the maximum capable performance. The Hurricane and Spitfire were without doubt significantly more advanced than any previous British fighters. The Hurricane was a conventionally built aircraft, made of canvas stretched across a wooden frame. Its maximum speed was 330 m.p.h. – well below that of its German rival. But one of the advantages of the Hurricane was that it was easy to build and so could be produced in massive numbers. The Spitfire, by contrast, was made of stretched metal panels on an all-metal frame, which meant it was a much more complex machine to produce. Designed by R.J. Mitchell, its bird-like elliptical wings made it one of the great classics of aeronautics. The models used in the Battle of Britain could fly at 360 m.p.h., about the same speed as the Messerschmitt Me-109s, but the Spitfires were highly manoeuvrable, especially at altitude, and their pilots had better all-round vision out of the clear cockpit. Tens of thousands of words have been written arguing that the Spitfire was superior to the Hurricane, or vice versa. The truth is that the RAF needed both aircraft to fight the Battle of Britain. Often the Hurricanes would be directed at the slower-moving bombers and the Spitfires would be sent to attack the fast fighter escorts. So without the large numbers of the Hurricanes and the superb quality of the Spitfires, the outcome of the Battle of Britain would definitely have been very different.[16]

The RAF pilots were good, but they lacked the combat experience of their Luftwaffe rivals. And the German fighters were organised around the *Schwarm* – or finger-four – formation, which could easily break down into two leading planes each supported by a wingman. This gave the Luftwaffe tremendous flexibility in combat. Although the need for flexibility had been a lesson learned by both sides in First World War aerial combat, in Britain it had been lost somewhere in the inter-war years. Most RAF commanders still insisted on rigidly following the *Fighter Area Attack Manual*, which called for squadrons to fly in four flights of three aircraft each, in a tight V formation. Many pilots spent much of their time ensuring that they were flying in correct formation, rather than scanning the skies searching out enemy aircraft. Furthermore, RAF tactics were based upon the assumption that British fighters would be attacking bombers flying alone from bases in Germany. As soon as France and Belgium fell, the Luftwaffe transported its fighters to air-bases only a few miles from the Channel coast, so they could escort the bombers over England. The RAF proved slow to adapt to the changed conditions of the actual combat they now faced. 'We did everything wrong that we could possibly do wrong,' said one fighter ace many years later. Rigidly adhering to outdated flying regulations cost the lives of many RAF pilots in the battle that summer.[17]

So, the RAF had some important advantages in intelligence, in command structure and in technology over the Luftwaffe. And most of the Luftwaffe effort in the war so far had been in ground support, attacking tactical targets set by the advancing army. This had proved fantastically effective in the invasions of Poland and France. But the Luftwaffe had not engaged in a bombing offensive before, let alone an offensive hundreds of miles from its bases across a sea. Nor did it have any four-engined bombers like those that would serve RAF Bomber Command so well later in the war. The Heinkel and Dornier bombers were good aircraft, but their bomb loads were small in comparison with the later British bombers. But the scene was now set for an intense aerial battle in which each side enjoyed some advantages over its enemy but suffered from disadvantages too.

The Battle of Britain was, by its nature, a defensive operation. But even at this stage of the war Churchill's restless mind constantly sought out opportunities for offensive action. Italy had declared war on 10 June, and with France now defeated,

Mussolini threatened British authority in the Mediterranean and Egypt. A huge Italian army of a quarter of a million men was assembling in Libya, near the Egyptian border. The commanding officer of British and imperial forces in the Middle East was General Archibald Wavell. He was in command of only some fifty thousand men in Egypt. So Churchill and the War Cabinet agreed during July to reinforce the Middle East garrison by sending shipments of light tanks to Egypt with a view to mounting a pre-emptive strike against the Italians. It was a brave move at a time when Britain itself was facing the threat of invasion. But it was typical of Churchill's belief in the need to seek out offensive measures. But the Italians were not yet ready to go on the attack in North Africa. Wavell was recalled to London for a series of meetings with Churchill. Wavell was an unusual senior officer, something of a poet and an academic, but socially timid, and he became tongue-tied in front of Churchill. Even worse, he did not stand up to the Prime Minister, who constantly requested information and wanted to debate every aspect of a military campaign. Churchill thought the worse of him for this, and their meetings were disastrous. He even considered replacing Wavell. But the situation at home was too serious and it was several months before the war in North Africa could get under way. It had to take a lower priority.

On 10 July, the first phase of the Battle of Britain began. The Luftwaffe's dive-bombers launched attacks on shipping in the Channel. They were led by the Junkers Ju-87, the Stuka. The siren that wailed as the plane went into its dive, which had caused such terror among troops and civilians across Poland and France, was now intended to have the same effect on Britain's mariners. But the attacks were fiercely defended by the young pilots of the RAF, whose fighters were faster and easily outperformed the dive-bombers. The Stukas began to suffer serious losses. On the first day of this phase of the battle, over twenty German aircraft were damaged with the loss of twenty-three aircrew. The Stukas soon began to be known by their German crews as 'flying coffins'.

Poor weather over the Channel caused something of a respite. But on 25 July visibility improved and the Luftwaffe mounted a further series of attacks on British shipping. These were intended to draw out more fighters, but Dowding was reluctant to respond as he wanted to maintain his strength for what he rightly saw as the tougher tests to come. Inspired by their war leader, the British people gritted their teeth in anticipation of the struggle ahead.

Meanwhile, in the Channel ports of France and Belgium, the Germans were busy assembling Rhine barges and other flat bottom boats that could be used to land men and machines on the English coast. During July, RAF aerial reconnaissance flights brought back alarming evidence that huge numbers, up to about two thousand, of these barges were being assembled. At one point the photo-interpretation officer who was counting the barges threw down his equipment and said: 'We don't want these. They'd better give us rifles.'[18] They looked threatening at the time. But four years later the Allies would find that their landings in northern France were largely dictated by the availability of the right sort of specially built landing craft. The Germans were already learning how important landing craft were and so rapidly improvised whatever suitable vessels could be found for Operation Sealion.

There's no doubt that at this point the threat of invasion was putting Churchill under great strain. The line between cajoling and pushing his military chiefs, on the one hand, and bullying and threatening them, on the other, was always a fine one. At his best, Churchill knew how far he could push people without alienating them. But under pressure he could go too far. Such was the case at Chequers, the Prime Minister's weekend residence in Buckinghamshire, on 26 July, when Churchill's frustration with his military chiefs spilt over into direct rudeness. At dinner he questioned General Sir James Marshall-Cornwall about the readiness of his divisions that were stationed on the Welsh Marches. When he discovered that some of the information he had been given in a paper by the CIGS Sir John Dill was less than adequate, Churchill flew into a rage and threw the report at Dill, demanding that it be checked and returned to him the following day. After an awkward silence, Churchill turned to 'the Prof', Frederick Lindemann, who was a regular at the Chequers dinner table, and asked him what news he had to tell him. To the amazement of everyone at the table, Lindemann produced a Mills hand-grenade and proceeded to explain why it was an inefficient weapon. The Prof announced that he had designed a far more efficient alternative. Churchill was delighted, and with almost boyish glee he told Dill to scrap the Mills grenade immediately and introduce the Lindemann grenade. Dill, no doubt spluttering through his soup, announced that contracts had already been placed for millions of the Mills bombs in both Britain and America, and it was impossible to cancel them. Churchill must have known this was true, but he continued to berate Dill. The Prime Minister ended the evening after a further interrogation of the generals by muttering peevishly, 'You soldiers are all alike; you have no imagination.' The tirade seemed designed to humiliate the military men in front of Churchill's civilian cohorts. The military chiefs probably forgave him, understanding the pressure he was under, but this was certainly the war leader at his most infantile, on his worst behaviour and displaying an ill-judged outburst to men who were themselves under great strain. General Marshall-Cornwall later described the evening as 'The Mad Hatter's Dinner Party'.[19]

At the end of July, Hitler held a military conference at his retreat in the mountains above Berchtesgaden to review plans for the invasion of Britain. Admiral Raeder once again called for a postponement to May 1941. Hitler was not happy but agreed to push back Operation Sealion to mid-September. Everything hinged on Goering and the success of the Luftwaffe. So, in early August, Goering decided to change his tactics. The thirteenth of August was set for *Adlertag*, 'Eagle Day'. This marked the beginning of the next phase of the Battle of Britain. Now the focus would shift from attacking Channel shipping to an attempt to destroy the RAF either in the air or by bombing its air-fields. This was a gamble by Goering, but he was confident that he would destroy the RAF and leave the British begging to surrender.

In the twenty-four hours before Eagle Day, the Luftwaffe attacked several of the radar towers along the British coast. The towers stood some 360 feet tall and were easy targets to iden-tify. Two radar installations in Kent were attacked but survived. The station at Pevensey was hit hard and several operators were killed or wounded. At Rye in East Sussex and at Ventnor on the Isle of Wight there was severe damage. The Luftwaffe had identified the 'eyes' of the RAF defence system. But in a remarkable blunder by the Germans, the radar towers were not consistently targeted again. Within days, they were all back in full working order.

On Eagle Day itself, Goering sent a command to each unit: 'Within a short period you will wipe the British Air Force from the sky. Heil Hitler.' But the day began badly for the Luftwaffe. With thick cloud cover, Goering gave the order to recall the first assault wave. The fighters got the message and returned, but the bombers did not and carried on unescorted – to a mauling over southern England. Later in the day, however, as the skies cleared, a force of some three hundred planes headed off to strike at the RAF airfields. By the end of the day, the Luftwaffe had lost 45 planes in action. The RAF had lost 13 fighters in the air and 47 aircraft on the ground – but only one of those was a fighter.

A major failing in the Luftwaffe was its inability to assess the damage it was causing the RAF accurately. Intelligence officers too willingly believed the daredevil claims of their pilots and repeatedly overestimated the number of 'kills'; consequently, they underestimated the number of planes the RAF could call upon. On the eve of Eagle Day, Luftwaffe intelligence officers estimated the RAF had 450 fighters left. In fact, the number was well above 600. And the Luftwaffe totally miscalculated the rate at which British factories were pouring out replacement aircraft. By late August, the Germans calculated that their enemy had barely 300 operational aircraft. In fact, the RAF had about 700 available fighters. Although Goering might have been delighted to hear the figures given by his intelligence officers, the pilots who flew daily over Britain grew to mistrust them, and this was appalling for morale.

As the battle continued to rage, with regular raids at dawn and dusk, Dowding's principal problem was not the numbers of aircraft but the loss of skilled pilots. A week after Eagle Day, four out of five squadron commanders in Fighter Command had been killed or wounded, or were resting after continuous combat. During August, Dowding reluctantly had to reduce the operational training period to two weeks. Before the war, it had been six months. So, at the height of the Battle of Britain, novice pilots with only a few flying hours under their belts were being thrown into the thick of the action.

The Hurricane and Spitfire pilots and their ground crews were on constant stand-by from half an hour before dawn, about 4.30 a.m., until around 9.00 p.m. The one feature nearly all of the pilots who survived the summer remembered was a feeling of constant tiredness. No doubt the continuous strain of waiting for the order to scramble added further to this. Most pilots lounged around all day, playing cards or chess. Some were able to read a little, others would doze, others listened to gramophone records. When the order to scramble came they would race to their aircraft. Often the ground crew had already started the engine. Quickly kitted up with parachute, Mae West life-jacket, headgear and linked into the radio and oxygen supplies, the pilot would then throw the throttle forward and the aircraft rolled across the grass and into the air. Every second counted. On really intense days pilots would fly three, four or even five separate sorties. Between each one the hard-working ground crew had to service and refuel the aircraft, make any instant repairs and rearm the four guns in each wing, leaving the plane totally ready for the next scramble at a moment's notice.

All the dogfights took place in full view of the civilian population below. Factory workers, farmers, mothers at home and children at school would all strain their necks and look up at the vapour trails criss-crossing the blue skies above. When planes came crashing to the ground they were quickly surrounded by the Home Guard and any enemy pilots who survived were marched off to the local police or the military, and into captivity. Sometimes even RAF pilots were pursued by over-keen crowds. Squadron Leader James Nicolson was the only fighter pilot to win a VC in the Battle of Britain. But when he crash-landed his plane, badly burned on his hands and face, he was fired on by a trigger-happy member of the Home Guard. And, of course, for young boys nothing could beat the cachet of finding and keeping a piece of a downed aircraft. To many on the ground below, the British fighter pilots became instant heroes as they fought gladiatorial battles against heavy odds.

Throughout August, Churchill busied himself with preparations for the expected invasion. He inspected defensive positions along the beaches where it might take place; he issued instructions about the roles of the Home Guard, the police and the fire brigade; and he approved plans for the defence of central London in the event of a German assault by parachute commandos. And every day he received reports on the numbers of German planes shot down, of British aircraft and pilots lost, and of the big increase in output from Britain's aircraft factories. He visited 11 Group's Operations Room in Uxbridge on several occasions. On 16 August, he was particularly moved watching all the markers being moved forward on the big table map as each new wave of German aircraft approached England and the markers of each of the RAF fighter squadrons as they were scrambled to intercept them. When he left in his official car, he turned to General Ismay and said: 'Don't speak to me; I have never been so moved.' After about five minutes of silence, he turned again to Ismay and uttered the words, 'Never in the field of human conflict has so much been owed by so many to so few.' Ismay was touched by the phrase that would soon reach a far wider audience.[20]

In the second half of August, the air battles grew more and more desperate. On Sunday the 18th, which has since been called the 'hardest' day, wave after wave of German bombers came over to attack the RAF airfields along with the aircraft factories. A vast force of 108 bombers escorted by several hundred fighters went for the fighter base at Kenley. Squadrons from across 11 Group were scrambled, and as further waves of bombers approached reserves from 12 Group in East Anglia, led by Air Vice-Marshal Trafford Leigh-Mallory, were called in to help. At the end of the day, the tally recorded 69 Luftwaffe losses with another 31 aircraft badly damaged. The RAF had lost 63 fighters with another 62 damaged. However, each side claimed it had inflicted much more severe losses on its adversary. On the Luftwaffe side, these inflated figures were believed and the crews were told they were winning the battle. In Britain, the simple fact was that the RAF knew it could not go on losing this many planes. At this rate, the RAF would cease to exist in a few weeks.

On 20 August Churchill addressed the House of Commons, speaking for nearly an hour. He said that this war had not seen the 'prodigious slaughter' of the First World War, but was instead 'a conflict of strategy, of organisation, of technical apparatus, of science, mechanics and morale . . . [and] our science is definitely ahead of theirs'. Then he spoke of the RAF pilots who had become the heroes of the hour, 'whose brilliant actions we see with our own eyes day after day'. He said that everyone's 'gratitude . . . goes out to the British airmen who, undaunted by odds, unwearied in their constant challenge and mortal danger, are turning the tide of war by their prowess and by their devotion'. Even though the Battle of Britain was by no means won by this point, Churchill reworked the words he had said in the car to Ismay just a few days before, proclaiming: 'Never in the history of human conflict was so much owed by so many to so few.' Churchill ended by announcing that he had at last won US approval for the loan of the fifty destroyers in the 'destroyers for bases' deal. He said this would bring together the 'two great organisations of the English-speaking democracies, the British Empire and the United States' which he did not view 'with any misgivings'. He concluded by saying: 'I could not stop it if I wished; no one can stop it. Like the Mississippi, it just keeps rolling along. Let it roll. Let it roll on full flood, inexorable, irresistible, benignant, to broader lands and better days.' As he sat down the House rose to its feet, with members cheering. Then Churchill returned to Downing Street and his private secretary recorded that he sang 'Ol' Man River', out of tune, all the way back in the car.[21]

Churchill was carrying out one of his great tasks as war leader. He was rallying the nation and stirring the soul of almost everyone who heard or read the speech. And internationally, he was beginning to persuade people that Britain was not down and out, was not staring at defeat as France had done in the face of Nazi aggression, but had the determination to fight on. Also, he had provided a nickname that would stick to the RAF pilots on whom the nation's survival depended: 'the Few'. One friend wrote that his words would 'live as long as words are spoken and remembered'.[22] Over time, the reference to the Few became the enduring memory not only of Churchill's speech but of the months that became known as 'Spitfire Summer'. But that summer was by no means over yet.

On 26 August, there were three major assaults. The strains within Fighter Command were beginning to show, and at one point during the day 12 Group did not come to 11 Group's aid when all its squadrons had been scrambled. As a consequence, the airfield at Debden was left unprotected and suffered a heavy bombing raid. Leigh-Mallory, the commander of 12 Group, was beginning to take a different view from Park as to the best strategy for defending Britain. He came to believe that his fighters were most effective when they formed mass formations of three or more squadrons. On the other hand, Park thought it best to attack at squadron level, keeping other squadrons back in order to respond to later assaults and to prevent all his forces being drawn into action too early. The argument became known as the 'Big Wing' controversy and it began to split the unified ethos of Fighter Command.

In early September, the Battle of Britain entered its crucial phase. The Luftwaffe was now focusing its attacks on the RAF airfields in the south-east of England in a bid to break the fighter defence of London. On 3 September, the Luftwaffe and the RAF each lost sixteen aircraft. Two days later, the RAF lost twenty-two and the Luftwaffe twenty-one. True, if a British pilot baled out and survived, he could be patched up and back in action within days – in some famous cases pilots were back flying within hours of being shot down. Whereas Luftwaffe crews who survived being shot down were usually captured, taken prisoner and represented a total loss to the German war effort. But in a two-week period the RAF lost 103 pilots killed and 128 badly wounded, with 466 Spitfires and Hurricanes lost or seriously damaged. To replace them were 260 eager but inexperienced pilots straight out of training. The RAF simply could not go on enduring this level of losses. It looked as though the Germans might soon win the Battle of Britain.

Hitler had ordered the Luftwaffe not to bomb British civilian centres, but on 24 August a small number of Heinkel He-111s became separated from their squadron and dropped their bombs, in error, on the suburbs of London. Nine civilians were killed. Churchill ordered Bomber Command to retaliate, and over the next few nights about eighty Wellingtons bombed Berlin. They caused only a tiny amount of physical damage but Berliners were outraged. Goering had said that no enemy bombers would ever fly over the capital of the Reich. Hitler was furious and lifted his ban on bombing British cities. To a small gathering in Berlin, he announced that Britain's cities would be 'razed to the ground'. The crowd roared with hysterical approval.

Goering called a conference of his Luftwaffe commanders in The Hague. They were still puzzled as to how the RAF was managing to put up so many fighter aircraft. According to their intelligence estimates, Fighter Command had hardly any fighters left. But almost every time they arrived over a target the German crews found dogged British pilots waiting for them. The German pilots mocked their intelligence officers by reporting back that 'the last fifty Spitfires' had again been waiting to intercept them. Although the Luftwaffe could endure more

losses than the RAF, the battle did not seem to be going well
from their perspective. Goering demanded another change in
tactics – the second in less than a month. With Hitler calling for
reprisals against British cities, Goering now committed the
blunder that ultimately lost the Battle of Britain for Germany.

At about 3.30 p.m. on Saturday 7 September, British radar
picked up the biggest force yet of German raiders about to cross
the Kent coast at Deal. This vast mass of bombers and fighters
filled an airspace of about eight hundred square miles. At
Bentley Priory, the plotting officers looked glum as further
reports came in from observers across the South-East. Park
scrambled several squadrons. But what happened next amazed
everyone. The raiders flew on, over the airfields they had been
bombing for days. They flew on across the green fields of
Kent, and over the suburbs of south-east London. They flew on
towards Goering's new target – the docks and factories of east
London. Once the first wave had dropped incendiaries and
created a huge inferno on the ground, the later waves came
in one after another to bomb and destroy the network of ware-
houses and docklands below. The London docks were the
heart of a vast trading empire. Enormous supplies of timber,
paint, rubber, flour and dozens of other commodities were set
ablaze. Firefighters struggled with fires that lit up the night sky.
The bombers continued to arrive until the early hours of the
following morning, and the 'all clear' was not sounded until
4.30 a.m.

Hundreds of acres of buildings burned to the ground. And in
the densely packed streets of the East End, where most of the
dockers lived, 448 civilians were killed and more than 1500
were injured. However, critically, the Luftwaffe had decided to
give up attacking the RAF airfields as this didn't seem to be
bringing victory. It was now going for civilian and industrial
targets to try to break the morale of the people. This was the

life-saver the RAF needed. Fighter Command simply could not have survived many more days of constant bombing of its airfields. Now it was the turn of the people of London.

During the course of this same Saturday, intelligence reports came in suggesting that the Germans were about to launch the invasion. Decrypted messages along with aerial photography seemed to show that barges in the French Channel ports were being readied for use. An emergency meeting of the Chiefs of Staff was called for 5.30 p.m. It was believed that the invasion might come on the following day. As the bombs started to fall on the docks and the East End, the Chiefs of Staff ordered all defence forces in the UK to 'stand by at immediate notice'. Just after eight o'clock that evening, General Headquarters, Home Forces, on their own initiative, sent out the code word 'Cromwell'. This was the sign that invasion was imminent. Across the country, the defence forces and the Home Guard went on to alert, and waited. But it was a false alarm. Hitler was not ready to invade.

Churchill visited several towns that had suffered from German bombing. He was very moved during a visit to Dover and Ramsgate in late August by the plight of those whose houses had been destroyed or badly damaged. He immediately said he would browbeat the Chancellor of the Exchequer into offering full compensation to the victims of Hitler's bombs. This policy was agreed by the War Cabinet in early September. The day after the bombing of the East End, Churchill visited the site of the heaviest devastation so far. Fires were still raging when he and his small entourage arrived. Whole rows of jerry-built houses had been reduced to piles of rubble. Tiny paper Union Jacks had been planted on some of these heaps of rubble. Today, one could imagine the survivors would turn on an authority figure and blame him for their lack of protection. But Churchill was literally mobbed. 'Good old Winnie!' people

called out. 'We thought you'd come and see us. We can take it.
Give it 'em back.' Churchill, who was always an emotional
man, broke down in the face of this defiant response. General
Ismay had the utmost difficulty escorting him through the
crowd, and he heard one old woman say: 'You see, he really
cares; he's crying.' Churchill remained in the Docklands until it
began to get dark, refusing to leave despite pleas for him to get
away. Then the air-raid sirens went off and the Luftwaffe's
bombers returned. Churchill's car got caught in a street
hemmed in by bomb damage, and a stick of incendiaries
landed only a short distance away. When the party finally
returned to Downing Street, Ismay was told off for taking such
risks with the Prime Minister's life. He responded angrily by
saying that anyone who thought they could control Churchill
on jaunts like this was welcome to try on the next occasion.[23]

That Sunday night, 412 more civilians were killed in the
bombing raids; and 370 on the following Monday. The Blitz on
London had begun in earnest. On 11 September, the RAF lost
thirty-one fighters. It was clear that the Battle of Britain was
reaching a climax. Four days later, the Luftwaffe attacked in
two large waves. Park's crews barely had time to refuel and
rearm between the raids. But Leigh-Mallory's Big Wing came in
from the north and caused serious havoc among the German
bombers.

The Prime Minister chose this Sunday to visit Park's 11
Group headquarters at Uxbridge once again. Churchill's wartime
memoir–history makes much of the visit. He describes sitting
on the top level, looking down on the plotting table below,
where twenty young men and women were moving the
markers representing enemy formations. It was like a 'small
theatre', where he and Clementine had seats in the 'Dress Circle'.
He noticed the numbers of Park's squadrons written on the
blackboards as 'Standing By', 'In Action' or 'Returning Home'.

Alongside were the numbers of Leigh-Mallory's squadrons as they came in to support. As the battle unfolded, more and more German formations moved across the big table top map and Park 'in a calm, low monotone' ordered more squadrons to scramble. Everything Churchill witnessed was at the peak of efficiency in this 'elaborate instrument of war'. But there was no doubting the intensity of the drama unfolding in front of everyone. Late in the afternoon, Churchill went over to a tense-looking Park and asked: 'What other reserves have we?' Park came back with the curt but chilling reply: 'There are none.' Churchill looked grave. He later wrote: 'The odds were great; our margins small; the stakes infinite.' Five minutes passed. Then the markers were moved slowly back across the big table. The Germans were returning home. As Churchill emerged from his underground theatre the 'all clear' was sounding. When he got back to Chequers, he was told that 183 enemy planes had been shot down for the loss of just 40 British aircraft. (The actual figures later verified for that day were 56 German losses and 27 British.) Churchill later wrote that this day was the 'culminating date' of the Battle of Britain, which was 'one of the decisive battles of the war and, like the battle of Waterloo, it was on a Sunday'. The fifteenth of September has been commemorated as Battle of Britain Day ever since.[24]

At the time, neither Churchill nor Park knew that when the Luftwaffe crews landed the German pilots reported with astonishment the news of the Big Wing attack on them from the north. Believing the RAF had almost no fighters left, they could not understand where these squadrons of Spitfires kept coming from. Dispirited, they counted their losses. Losing 56 planes and their crews was a serious blow. This rate of attrition could not go on. Two days later, realising that he had failed to win mastery of the skies, Hitler issued a secret order postponing Operation Sealion. Within weeks, he had started to plan his invasion of Russia and the next dramatic extension of the war.

Daytime raids on airfields and cities continued sporadically into October. But from now the bombing of Britain moved into a new phase known generically as the Blitz. The Luftwaffe preferred night-time raids on the cities and industrial centres. London was blitzed, with only one night's respite, for seventy-six consecutive nights. The toll of cities across Britain being heavily bombed increased weekly – Portsmouth, Bristol, Plymouth, Birmingham and Liverpool all joined the list. The worst raid came on the night of 14 November, when Coventry was bombed. 554 civilians were killed and 1200 injured. There were so many corpses that they had to be lined up in rows for burial in mass graves. During the Blitz, the Second World War became the People's War. In total, about 43,000 civilians were killed, well over 150,000 were injured, and a quarter of a million were made homeless

But Churchill was right: 15 September had been the decisive day. And the Battle of Britain had been won. Not in the conventional sense, by destroying the enemy: the RAF had lost 1173 aircraft, with 510 pilots killed, while the Luftwaffe had lost 1733 aircraft, with over 3300 airmen killed, wounded or taken prisoner. But the Luftwaffe was still a major force, and would remain so for at least three more years. But it was a victory for Britain because Hitler had set out to destroy the RAF, to win mastery of the skies over England and then, possibly, to invade. And he had failed. As the invasion barges were quietly redeployed and sent back to their Rhineland owners, the RAF still ruled the skies over Britain. The British people had not been brought to their knees. And, in the longer spectrum of the war, Britain had survived as a base from where, when the Allies were ready, the invasion of Europe and the final defeat of Hitler's Third Reich could be launched.

Much has been written about the spirit that unified first Londoners and then the whole nation in the face of Hitler's Blitz. The traditional view that 'we was all one' and that divisions were put aside for the duration has been severely challenged by historians in recent decades.[25] There is no doubt that occasionally there was panic. At times local authorities failed miserably in their attempts to manage the chaos that befell their residents, and petty-officialdom left people frustrated, angry and vulnerable. At times people flooded out of the cities in fright, even when they were told not to. At times looting of bombed houses took place before their owners had returned from the air-raid shelters. People in Britain displayed all the traits that people everywhere exhibit during bombing. Fear becomes terror, which becomes anger, which becomes exhaustion, which extends into a burning desire for revenge. But there is also no doubt that Churchill had given the nation a fillip throughout this formidable summer. His visits to the sites of bomb attacks cheered the victims in a way that is difficult to imagine in today's more cynical age.

During the war years, British people made about twenty million visits to the cinema each week. It was a hugely popular medium for entertainment and before the main movie came the newsreel, twice-weekly compilations of news stories that were part of every cinema show. The newsreels were quite unlike today's television news. Reliant upon 35mm film that had to be processed, developed and edited, they were always a few days out of date. Nevertheless, some of the newsreels of this period capture the popular attitude towards Churchill very clearly. Unlike the heavily stage-managed performances seen in the newsreels of Hitler and Mussolini, Churchill was usually shown visiting bomb sites or touring factories. He always wore a hat which he would wave at the crowds or put on the end of his walking stick to wave in the air if the crowds around him could not see him clearly. He always had his characteristic cigar as a prop. And he would regularly hold up his fingers in the V-for-victory sign. He nearly always looked defiant and even when visibly moved by what he witnessed his bulldog spirit seemed to come across. These were not scenes directed by some clever news producer. This was not spin managed by the Ministry of Information or the Cabinet Office. It was just Churchill being himself. He would have made all of these visits even if the newsreel cameras had not been there to film him. And just as his spontaneous visits to victims of air raids proved popular to those who saw him in person, so the newsreel scenes projected to a much larger national audience the same message of encouragement and inspiration.

But it is for his words that Churchill is best remembered from the summer of 1940. Words had been vital to him since the beginning of his adult life. And from him came a torrent of words either performed as hundreds of speeches or dictated as dozens of books, thousands of articles, and innumerable memos, minutes and official state papers. The historian David Cannadine describes his extraordinary career as 'one sustained, brightly lit and scarcely interrupted monologue'.[26] For much of his life he had been respected for his oratory, but people had still been suspicious of him. Churchill could always find a brilliant phrase, but was he right in what he said? In the 1920s, Chamberlain had said his 'speeches are extraordinarily brilliant and men flock to hear them . . . The best show in London, they say. But so far as I can judge, they think of it as a show, and are not prepared at present to trust his character, still less his judgement.'[27] But during the summer of 1940 Churchill was perfectly attuned to the public mood. His black-and-white way of seeing things enabled him to present the struggle as a noble one between victory and defeat, freedom and tyranny, civilisation and barbarism. The vast majority of adult Britons listened to his broadcasts in 1940, nearly three out of four of the population.[28] Many people seem to have felt that Churchill was expressing their own feelings but in a way they could not. His speeches became weapons in the war against Hitler. When very little stood between the British people and the Luftwaffe, invasion and the great likelihood of defeat, Churchill's defiance and his sense of history gave Britons a feeling of pride that they were living through hours that were as vital as when the Spanish Armada had sailed or when Napoleon had threatened. His speeches hardened their resolve that Britain could take it – and hit back again. Humbly, Churchill later said that it fell to him to express the will of the nation but it was the people who had 'the lion heart. I had the luck to be called upon to give the roar.'[29] But this is to play down his role. A New Zealand editor wrote to the BBC, saying, 'A speech made by Mr Churchill is as good as a new battleship.'[30] The great American journalist Ed Murrow, who reported from London at the time, summed it up best when he observed that Churchill 'mobilized the English language and sent it into battle'.[31]

By September 1940, Churchill had become the unchallenged leader of the people of Britain. A Gallup Poll reported an extraordinarily high 88 per cent approval rating.[32] Doubts and scepticism about whether he was the right man for the job had been forgotten. All the fears of Conservatives that he could not be trusted, that he was unreliable and too much of a loose cannon, were put aside. When he had made his 'finest hour' speech in June, observers had noted that it was mostly the Labour benches that cheered him in the Commons. Now, the Conservatives were also behind him to a man. Never has the popularity of a prime minister risen so quickly and so dramatically as in the four historic months of the summer of 1940. It was an extraordinary turnaround. At the height of the battle, Churchill said to his dinner guests one evening that he was puzzled as to why he was so popular. Since he had come to power, 'everything had gone wrong, and he had nothing but disasters to announce' and his platform was only 'blood, sweat and tears'.[33] But this suited the mood of the people. Although no one knew for sure that the threat of invasion had been lifted as the Battle of Britain merged into the Blitz, Churchill had successfully led the nation through its darkest hour. But the task of going forward to victory would be long and arduous. Churchill had yet to prove that he had the right formula, and the right people around him, to meet this challenge.

5

The Wizard War

In the early hours of 5 November 1939, a parcel was left on the window ledge of the British Consulate in Oslo, in what was then still neutral Norway. Signed simply 'A German scientist who wishes you well', the parcel contained several typed pages which appeared to relate details of the latest German scientific research. This included information about radar equipment, new fuses for bombs and shells, the progress of the dive-bomber, the development of rocket technology, details of a large experimental establishment at Peenemünde on the Baltic, and a description of a radar aid to guide bombers at night, called 'Y-Geraet' (or 'Y-Apparatus'). When it was passed on to Scientific Intelligence in London, it was received with much scepticism and it was thought that the document was a plant to confuse or mislead British scientists. It was ordered that copies of the report should be destroyed. In fact, as the war unfolded, one after another of the details in the document known as the 'Oslo Report' came real. The Report revealed that the war would be fought between scientists as much as between soldiers, sailors and airmen. Churchill called this 'the Wizard War', and the scientists and innovators he called up in the service of Britain were essential members of his War Lab.

Probably ever since human beings first struck an animal with a stone, they have tried to improve the sharpness of the flint or the destructiveness of the stone. Ever since men first used metal weapons to strike at their adversaries, they have tried to improve the efficiency and power of their arms. And certainly ever since firearms were first introduced in the late Middle Ages, gunsmiths have tried to improve accuracy, the rate of fire and the general destructive capability of their weapons. Leonardo da Vinci is often remembered for the genius of his art, but he actually devoted more of his life to the improvement of military machines for his patrons and the invention of new devices for destroying the enemy. In other words, the application of science has never been far from the development of the technology of war.

But around the time of the Industrial Revolution, the military and the world of science, the soldier and the scientist, became separated. As the profession of arms became a full-time professional calling, it developed its own mores, customs and practices. Soldiers became convinced that they, and they alone, understood the business of war and needed to get on with it without interference from outsiders who lacked their professional expertise. During the nineteenth century, soldiering became an inherently conservative profession. Officers whose decisions could directly result in the death of their men, and generals whose strategies could end in massive loss of life or national humiliation, did not want to experiment with new methods and ideas that might lead to even greater losses on the battlefield. Around these attitudes grew up the rituals of officer life associated with the world of soldiering in many

Western countries – the customs of the regimental mess and the subtle but rigid social hierarchies of different regiments. Soldiering, at least as far as officers were concerned, generated a socially exclusive lifestyle based on tradition and the repetition of the same processes over and over again. In Britain, for example, it now seems astonishing how many admirals were reluctant to give up sail and embrace the transition to steamships. In the army, many generals continued to champion the value of cavalry units even as the world became dominated by the internal combustion engine. And with this conservative view on how to fight wars (it is often said that each new war is approached using the methods of the last), the unchallenged acceptance of command and authority became an inherent and vital feature of the military mind. While science was intended to challenge, to question and to change the world, the armies and navies of most developed nations chose to freeze their thinking, in well-established and proven systems. As Solly Zuckerman, an important player in this story, put it: 'Where it is the habit of the scientist to question, it is that of the soldier to obey.'[1]

This was the world that Churchill entered first at Sandhurst and then in the 4th Hussars, one of the elite cavalry regiments of the British Army. But as we have seen, Churchill was no model young cavalry subaltern. He read books to stretch his mind. Instead of fox-hunting during his long vacations, he travelled to distant military conflicts to get a taste of the action. Then he wrote about these experiences and even went so far as to question the decisions of his senior officers. Although Churchill entered the British Army at the peak of its Victorian mode of thinking, and although he enthusiastically took part in its last great cavalry charge, he did not possess the frame of mind required to progress to the top of the military system. His mind was too restless, and he was far too ambitious to settle for a career of slow and gradual promotion through the military hierarchy.

When he returned to the military world as First Lord of the Admiralty in 1911, Churchill was much more associated with radical thinking, so it was not surprising that his fertile mind found new ideas appealing. Hence his enthusiasm for the Royal Naval Air Service, his vigorous support for the transition from coal- to oil-fired turbines, and when war came his encouragement for a variety of new technologies – from Q-ships to the tank to the code-breaking carried out in Room 40. Churchill was deeply upset and offended when his idea for a mechanical device to cross the barbed wire of no man's land and penetrate the enemy's trenches (what became the tank) was rejected by the army as 'not likely to lead to success'. His anger is still there in the pages of *The World Crisis*, written some ten years later.[2] When it came to his spell as Minister of Munitions in the last fifteen months of the war, again he was eager to find new systems and new ways of operating that would increase output and efficiency. And in the post-war era he sought out more new technologies and new strategies to devise a role for the RAF to act as a form of imperial police force, and to establish new accords with old enemies, as in the Irish Treaty negotiations. Churchill was always up for a challenge, and nothing was sacred to him in the traditional world of military thinking. As a young man, he had been deeply opposed to stuffy, closed-minded, conservative military commanders. He had come across them in India, the Sudan, in the Boer War and at the Admiralty. And in many of his books he had written of the need for vigorous and inventive thinking. 'Nearly all the battles which are regarded as masterpieces of the military art . . . have been battles of manoeuvre in which very often the enemy has found himself defeated by some novel expedient or device, some queer, swift, unexpected thrust or stratagem.'[3] In a sense, this statement sums up his quest for new and radical solutions to military problems. By the 1940s, this inevitably meant some sort of marriage between the scientist and the soldier.

Churchill, of course, was the first to accept that he himself understood little about science and almost nothing about mathematics – the subject he had found so difficult to master at school, while his failures in maths exams had nearly prevented him from qualifying for the army. So it was that during the late 1920s Churchill developed a friendship and deep respect for Professor Frederick Lindemann, known always as simply 'the Prof'. Lindemann was in many ways the most unlikely member of Churchill's court at Chartwell. He was a vegetarian, and a non-smoking teetotaller! He always looked the same, dressing in a dark formal suit. He wore evening dress at dinner. And he always took a bowler hat and umbrella when going out, whatever the weather. He was an eccentric who seemed to enjoy annoying people. Whenever he crossed the road, he never stopped to look, he would just step off the pavement, waving his umbrella, and charge through the traffic (there was less then than today). In conversation, he liked to be blunt and provocative. When someone at Churchill's dinner table at Chartwell before the war asked him for a definition of morality, he replied: 'I define a moral action as one that brings advantage to my friends.' Clementine whispered to the person who had asked the question: 'Doesn't the Prof sometimes say dreadful things?'[4]

But Lindemann was a physicist of great standing and international renown. He had known Einstein before the First World War while carrying out research in Berlin. He had joined the Royal Aircraft Establishment at Farnborough in 1915 where he had carried out important work, including finding a way for pilots to pull out of a spin, which in those days was nearly always fatal. He is supposed to have learned how to fly in order to carry out tests to see if his mathematical calculations about this actually worked in practice. They did. In 1919, he went to Oxford as Professor of Experimental Philosophy. At that time, Oxford science was way behind that of Cambridge, and Lindemann gave a huge boost to the work of the Clarendon Laboratory, which had been severely neglected.

Churchill later described Lindemann's value as one who could explain to him 'in lucid, homely terms what the [scientific] issues were'. Lindemann's vast and wide-ranging scientific knowledge and his natural self-assurance enabled him to sum up almost any scientific question for Churchill, whether it be the potential power of the atom or (as we have seen) problems with the standard-issue Mills grenade. 'There are only twenty-four hours in the day,' Churchill wrote, 'of which at least seven must be spent in sleep and three in eating and relaxation. Anyone in my position would have been ruined if he had attempted to dive into depths which not even a lifetime of study could plumb. What I had to grasp were the practical results.'[5] And this was what Lindemann provided him with. His memos to Churchill were usually just two pages long, double spaced in large print. He always tried to condense even the most complex and demanding scientific ideas into these bite-sized chunks for his boss's consumption.

When Churchill returned to the Admiralty in 1939, he asked his friend to come with him, and Lindemann left Oxford to do so. When Churchill became Prime Minister, the Prof went on to be one of the key players in his War Lab. He was made head of the newly created Prime Minister's Statistical Branch, with a tiny team of some six or seven economists and a scientist. It acted like an independent think-tank for Churchill and had a roving commission to dig into any aspect of wartime government and administration. Lindemann met with Churchill almost daily, advising him on scientific matters, military issues, logistical problems and even the economy. Most weekends, he joined Churchill and his entourage at Chequers. And he sometimes accompanied Churchill when the Prime Minister travelled abroad. Lindemann sent Churchill about two thousand memos during the war, equivalent to roughly one per day. His biographer said his role as scientific adviser to Churchill gave him 'power greater than that exercised by any scientist in history'.[6] Inevitably, with such a difficult and combative man, this would prove controversial. And, equally inevitably, with his prickly and vain personality, often coming across as pompous and stiff, there would be major fallings-out between Lindemann and other key scientists.

But Churchill was worried that the Whitehall establishment would be slow to respond to new and dramatic developments in science, and that good advice would take too long to percolate up to him. So he wanted Lindemann at his side not only to explain in a way he could understand what the new science might be, but also to advise him on what else was needed in the wizard war. The Statistical Branch prepared albums of tables and charts to illustrate at a glance the strength of military units in various theatres, and to keep a running record of shipping losses in the Atlantic. Churchill was proud of these albums and used to show them off to the King and, later, to President Roosevelt.[7]

Back in the early 1930s, Prime Minister Stanley Baldwin had said in Parliament: 'the bomber will always get through'.[8] By this, he meant that there was no effective means of defence against enemy bombers. In the age of appeasement, it became government policy simply to accept the inevitability that Britain would be bombed in a future conflict. Churchill and Lindemann did not accept this defeatist attitude and Lindemann wrote to *The Times* on 8 August 1934:

Sir, In the debate in the House of Commons on Monday on the proposed expansion of our Air Forces, it seemed to be taken for granted on all sides that there is, and can be, no defence against bombing aeroplanes . . . That there is at present no means of preventing hostile bombers . . . I believe to be true; that no method can be devised to safeguard great centres of population from such a fate appears to me to be profoundly improbable . . . To adopt a defeatist attitude in the face of such a threat is inexcusable until it has definitely been shown that all the resources of science and invention have been exhausted.[9]

Lindemann's letter acted like a wake-up call. Something had to be done to develop defensive means against the bombing of Britain. This was where the wizards came in.

A committee was set up to investigate how scientific and technical advances could aid the detection of enemy aircraft. Sir Henry Tizard was asked to chair it. Tizard has been called one of Britain's greatest defence scientists.[10] He had begun as a chemist but had given up pure research and moved across to find ways of applying scientific advances to practical problems. Tizard was an excellent chairman and was known for asking clear and brilliant questions. At this point he was the rector of Imperial College, London, to which he had given a great boost – just as Lindemann had improved the standing of Oxford's Clarendon Laboratory. In many ways, Tizard's career closely paralleled that of Lindemann. Indeed, they had known each other in Berlin before the First World War and had become friends. But they had very different views on how to bring science into the mainstream. Tizard committed himself to public administration and worked tirelessly on several government committees. Lindemann attached himself to Churchill and saw political alignment as the way forward. When Lindemann pressed the Air Ministry to set up a committee to explore the science of air defence and discovered that Tizard was already running such a committee, he took this as a great personal affront by both Tizard and the ministry. Instead of working together, Lindemann fell out with Tizard. And once he held a grudge, the Prof was not one to forgive.

In this context, significant developments were made in Britain in Radio Detection-Finding, radar. When Churchill joined the Air Defence Research Committee in 1935 he first learned about the early development of radar from Tizard and others. Radar's origins lay in the bizarre quest to find a 'death ray' that would send enough energy along a beam to destroy an enemy aircraft. In 1935, Robert Watson-Watt, the superintendent of the National Physical Laboratory's Radio Research Station at Slough, was asked to investigate this. He very quickly proved that the amount of energy needed was so vast that the death ray would remain firmly in the realms of science fiction rather than science fact. But Watson-Watt was one of the outspoken mavericks of British science in the 1930s, with a strong sense of how science could be used to transform the future. He observed that radio waves will bounce off an aircraft and after tests using the BBC's short-wave radio transmitters at Daventry, he devised a system for measuring the position of a flying object. The Air Ministry quickly grasped the potential of this invention to give an early warning of the approach of enemy raiders, an advantage that in the age of 'the bomber will always get through' would prove vital.

Air Marshal Hugh Dowding was at this time in charge of RAF Research and Development, and he was impressed. He agreed to put up ten thousand pounds, a substantial sum, of research funding for the device to be tested. In simple surroundings first at Orfordness on the Suffolk coast and then at nearby Bawdsey, Watson-Watt and his small team of scientists developed one of the great inventions of the twentieth century. They began to find ways of measuring first the distance, then the height and finally the bearing of aircraft using radio transmitters and a cathode-ray tube. Despite the low level of defence spending, radar (or RDF) developed rapidly under Watson-Watt and both the army and the navy began to show an interest in his experiments as well. As we have seen, RAF Fighter Command when Dowding took it over developed a plan for the organisation of the air defence of England based on the construction of the Chain Home radar system. But the principal achievement of Dowding and his team of scientists was to integrate the scientific side with the operational, to filter and process the information gleaned from the new science into an effective battle plan. Tizard played a vital role in this development during a series of tests at Biggin Hill airfield in Kent, in the late 1930s. He calculated the angle at which the faster fighters needed to be directed in order to intercept the slower bombers. This became known as the 'Tizzy angle' and was used until the 1960s, when computers took over such calculations. The whole process marked a subtle but important shift. Pilots, who had traditionally operated under their own rules of patrol and observation, were now scrambled and directed to the enemy by their controllers, who acted according to strict scientific principles in reading, interpreting and predicting the Luftwaffe's flight path. This has been described as being 'of crucial importance in the new service–scientist relationship'.[11] During the Battle of Britain, radar literally meant the difference between victory and defeat. It was not the only element that helped bring victory to the RAF. But without radar, defeat would have been certain.

In the run-up to war, several senior scientists realised the important role science would play in the future conflict. Fearing that the government was not doing enough to prepare for war, they drew up a list of seven thousand scientists, some working in universities, others in industry. Each scientist's name and area of expertise were entered on to a card index known as the 'Central Register'. By this simple method, physicists, engineers, chemists, biologists, astronomers and botanists were all carefully listed. So, for instance, when the Merchant Navy needed a specialist in maritime refrigeration, the right person could be quickly found. Although many senior officers in the army and navy were still sceptical about how these scientists would fit into military work, the Air Ministry was more welcoming, as it had been with the integration of radar into an operational defence plan. The RAF was the youngest of the three services and needed science in obvious ways to improve its performance in the air. So it is not surprising that this new breed of scientist, known as the 'boffin', should have been welcomed first by the nation's aviators. 'Boffin' was the affectionate term that came into popular use during the war to describe a scientist known for his inventiveness, his persistence and his ability to come up with weird and wonderful solutions to problems. (They were nearly all men, as women received little encouragement in the sciences at this time.) Sometimes they joined the military, often they remained civilians, operating through a range of government advisory committees. There was a lot of potential for tension here. Senior military officers were usually drawn from the gentry or the upper middle classes and were of a conservative disposition. The boffins usually did not enjoy much social status, came from a variety of backgrounds, and were often radical in their approach. Nevertheless, the boffins would bring much to the military over the next few years, and without doubt they contributed significantly to the Allied victory.

One of the greatest triumphs of science in the war was the deciphering of German codes. In the 1930s, the German military had developed a form of top-secret communication using the Enigma machine. This was an electro-mechanical typing machine which was able to encode every letter of every message via a series of rotor blades. The scrambled message was then sent as a conventional radio signal. The Enigma machines for the various parts of the military were slightly different in their configurations, but the basic operating principle was the same. The operators would reset the rotor blades at the back of the machine every twenty-four hours, so that messages would be encoded according to a different formula each day. The German military were convinced that their Enigma codes were secure because, although it was theoretically possible for the enemy to decipher a message, it was reckoned that it would take so long that by the time it had been done the rotor blades would have been reset and a new code created. The Germans therefore put total faith in their Enigma codes, and the High Command communicated regularly with the Wehrmacht and the Luftwaffe, issuing orders and receiving field reports filled with precise details about locations, the strength of units, casualties, operational plans, and so on.[12]

Before the war, Polish Intelligence had captured an Enigma instruction booklet and had even got its hands on an Enigma machine, and so had been able to crack the German code system. In the summer of 1939, French Intelligence and the British Secret Intelligence Service started to get interested in this. Just before the outbreak of war, the Government Code and Cypher School moved its staff of about one hundred cryptographers to Bletchley Park, a country-house estate to the north of London. Additional teams of mathematicians were then recruited to work there, many from nearby Cambridge University. After the defeat of Poland and then France, Bletchley became the principal Allied centre for code-breaking. The amount of work that went on here soon outgrew the mansion house and stables, and dozens of brick huts were built across the grounds of the estate. Within a few years, seven thousand men and women were working at Bletchley, with even more at a series of outstations around the country, from Dorset to northern Scotland. They included radio operators, mathematicians, decryptologists, interpreters, and hundreds of clerical support staff working on the central index, many of whom came from the Women's Royal Naval Service.[13]

Of all the many extraordinarily brilliant personalities who worked at Bletchley, Alan Turing was one of the most exceptional. A top-level mathematician and mechanical engineer, he was a fellow of King's College, Cambridge, and had written a pioneering paper on computable numbers before the war. He was only in his late twenties when he arrived at Bletchley, with boyish looks but a totally dishevelled and eccentric air. His trousers were often held up with an old tie, he had stopped shaving regularly, his hair was scruffy and he spoke with a stutter. He was painfully shy and developed the habit of working continuously for days at a time before collapsing in exhaustion. His military masters never really understood him. But he ran Bletchley's famous Hut 8 where he helped design the first 'bombes', huge electrical machines six-foot-by-seven, consisting of thirty rotating drums that ran through thousands of letter possibilities at high speed to find the correct match of plain text with encrypted letters. Out of these later in the war came Colossus, the world's first operational computer, which could run through the tens of millions of computations that were necessary to decode some of the most complex messages. Initially it took days to decode a signal, but Colossus reduced this to hours and ultimately to minutes. It was a stunning breakthrough and Turing went on to help found the post-war computer industry. Hounded for being gay, he committed suicide in 1954 by biting on a poisoned apple.

Although exceptional, Turing was only one of many eccentric and remarkable characters working as code-breakers at Bletchley Park. Gordon Welchman, another pipe-smoking, studious Cambridge mathematician, was one of the first to recognise the scale of the task that faced code-breakers who had the potential to listen in to the German high commanders talking to each other on a daily or even an hourly basis. When more code-breakers were needed, he drove off to Cambridge in his Austin 7 and rounded up former colleagues and students to join his team. Peter Twinn, an Oxford mathematician, found that he was resented by the old school of cryptologists, who were all classicists and suspicious of what these bright young mathematicians could contribute. Stuart Milner-Barry was an international chess champion who was recruited at the start of the war. Josh Cooper was one of the strangest. He had the peculiar habit of putting his right hand behind his head and stroking his left shoulder when he was thinking. He was also known for every now and again missing his chair when he went to sit down and landing up on the floor. Mostly the newcomers were young, in their mid-twenties, but there was also a smattering of older men who had been members of the Admiralty's Room 40 in the First World War. Nigel de Grey was one of the most celebrated Great War code-breakers. Frank Birch, another Room 40 veteran, had combined life as an academic with a career on the pantomime stage. Some of the newcomers were women, like Diana Russell Clarke who worked in the Hut 6 Machine Room. She was renowned for driving her Bentley sports car at high speed through the local country lanes. Linguists Phoebe Senyard and Barbara Abernethy were also among the earliest recruits.

Today, the work of Bletchley Park (or Station X, as it was known) is famous. There have been novels, movies and TV series about the place. But between 1939 and 1945 it was the most highly secret operation in the entire war effort. Very few people, including most of the staff at Bletchley themselves, had any idea how crucial this code-breaking work was. The recruits were placed in a small group and got on with their own tasks, totally unaware of what was happening in any other group. Most of the new arrivals were selected because they were thought to be 100 per cent reliable, but on arriving at Bletchley many were often still met by a security officer who would draw his pistol. Everyone was told that he or she must never breathe a word of anything they did or knew about to anyone else. If they did, they were told, the officer would personally come and shoot them. The secrets of Bletchley Park lived on long after the war. It was only in the mid-1970s that stories began to emerge about the code-breaking that went on there and its importance to the Allied victory.

The information gleaned from breaking the German codes and listening in to the Enigma communications between field units and their headquarters was known generically as 'Ultra'. It began to come on line within weeks of Churchill becoming Prime Minister. The excitement of reading deciphered messages direct from the enemy appealed greatly to Churchill, just as it had in the First World War. He really enjoyed the magic and the mystery of it.[14] Instead of reading summaries of the messages, he asked to receive the information raw, as it had been decoded and translated. He did not like the idea of it being watered down or distorted to give him a rosier picture than was really the case. Every day, wherever he was, a special box in faded yellow leather was delivered to Churchill by a messenger from the Secret Intelligence Service (SIS). In this box were the latest decrypts, the decoded messages. Only Churchill held the key to this box, on his keyring. The boxes came to him from the director of SIS, the man known as 'C' (later the inspiration for 'M' in the James Bond stories written by Ian Fleming, who worked in

Naval Intelligence during the war). 'C' was Colonel Stewart
Menzies, who ran SIS throughout the war and into the early
1950s. Desmond Morton coordinated the flow of Ultra from
Menzies, but it all went directly to Churchill, who returned it
after reading. To protect the source of this information, the code
name 'Boniface' was used, suggesting to anyone who might by
accident hear about it that it had been supplied by an agent, a
spy operating somewhere inside Germany. Even Churchill's
closest aides knew nothing about these decrypts. Only a tiny
number of senior ministers and the Chiefs of Staff and their
deputies, about thirty people in all, knew of the existence of this
highly valuable top-secret intelligence. Over the next few years,
the mass of information decoded from the Wehrmacht, the
Luftwaffe, the Abwehr (German Military Intelligence) and even
from the German police and railways revealed much about the
enemy's military intentions. The German Navy used an even
more sophisticated Enigma system with extra rotor blades,
adding immensely to the challenge of the code-breakers. Only
when these naval codes were finally broken in December 1942
did the Battle of the Atlantic begin to turn in Britain's favour. At
the time, Churchill called the code-breakers 'the geese that laid
the golden eggs and never cackled'.[15]

In September 1941, the Prime Minister made a personal visit
to Bletchley Park to inspect the 'geese' for himself. Even he was
surprised at the casual dress and eccentric behaviour he saw.
He is supposed to have said to Menzies: 'I know I told you to
leave no stone unturned to get staff but I didn't expect you to
take me literally.'[16] He made a short and emotional speech to
the code-breakers, telling them how important their work was.
But because so few people in senior government positions were
allowed to know about the work going on there, Bletchley Park
was constantly turned down when it put in bids for much-
needed extra staff and resources. By this time the volume of

work coming through required a major expansion. So, a month after his visit, Turing, Welchman and two other leading code-breakers wrote directly to Churchill, telling him: 'We think you should know that this work is being held up, and in some cases not being done at all, principally because we cannot get sufficient staff to deal with it.' They went on to appeal above the heads of their bosses direct to the Prime Minister: 'For months we have done everything that we possibly can through the normal channels and we despair of any early improvement without your intervention.' They asked for additional typists, more clerks, and for the removal of various bottlenecks. When he received the letter, Churchill sent a minute to General Ismay, saying: 'Make sure they have everything they want as extreme priority and report to me that this has been done.' Then he stamped the minute with the red 'Action This Day' sticker. Within a month, the essential expansion at Bletchley Park had begun. The Ministry of Works started erecting new buildings and a recruitment programme for two thousand extra staff was launched. Once again, Churchill's personal intervention had made the key difference.[17]

Before the Battle of Britain had got fully under way and the genius of radar had proved its worth, the boffins again came to the aid of the military. A young research scientist who had worked for Lindemann at Oxford, Dr R.V. Jones, warned the Prof that the Germans had developed a sophisticated system of beams to guide their bombers to their targets. The beams could be used by day or night and in any weather conditions. Lindemann reported this warning to Churchill, who instantly recognised its importance. Without waiting for the slow-moving bureaucracy to assess the situation, Churchill called an urgent meeting on 21 June in the Cabinet Room. The new Minister for Air, Sir Archibald Sinclair, and the Minister for Aircraft Production, Lord Beaverbrook, were there along with

Tizard, Lindemann, Watson-Watt and several senior RAF figures. Jones was summoned to attend the meeting but when he got into work that morning and found the message calling him to the Cabinet Office he thought it was a practical joke. As a consequence he was about half an hour late. Soon after he arrived, Churchill asked him to outline the position. Jones was only twenty-eight years old, but undaunted by the top brass that now confronted him he launched into an explanation of how for some time he had been picking up reports that the Germans had developed a secret weapon, a new system of night bombing, on which they placed great hopes. It seemed to be linked to the code word *Knickebein*, which somewhat mysteriously translated as 'crooked leg'. A bomber had crash-landed earlier in the year and the phrase '*Knickebein* beacon' had been found in some documents. Captured Luftwaffe crew were interrogated and revealed the use of special new radio equipment. More shot-down aircraft were searched and further references to *Knickebein* were found, including one linking it to a location in Cleves, in north-western Germany. Meanwhile, aerial photography had revealed the existence of several strange towers that did not look like conventional radio beacons. After Luftwaffe prisoners of war were overheard saying, 'They'll never find where it is,' Jones had the idea that maybe this new device was contained within the existing system that enabled a pilot to land at night or in bad weather by tuning in to a beam. Sure enough, this equipment proved to be far more sensitive than was needed for landing purposes, and when an RAF aircraft tried it out, the plane picked up a whole series of signals linking it to a beam. It was then that Jones realised these could be used not for landing, but for guiding bombers to their targets.

Churchill and the distinguished gathering of senior officials listened to Jones as he told his story. Churchill later wrote 'For twenty minutes or more he spoke in quiet tones, unrolling his chain of circumstantial evidence, the like of which for its convincing fascination was never surpassed by the tales of Sherlock Holmes.'[18] Then there was a discussion around the table. Some of those present were incredulous. They argued that such a system was unnecessary and asked why the Luftwaffe pilots did not simply navigate by the stars, as RAF crews were trained to do. But Churchill, on the advice of Lindemann and against that of Tizard, who was sceptical about the existence of the beams, was ready to accept that the Germans had devised such a system. As the argument continued, Churchill grew angry and banged his fist on the table.[19] He ordered that countermeasures were to be investigated as a matter of priority. Jones went back to his desk in the Air Ministry to coordinate what became known as the 'Battle of the Beams'.

Conventional wisdom had it that short-wave radio beams did not have the accuracy to guide aircraft over the distances involved, that they would disperse like the beam of a searchlight with the curvature of the earth. But Jones persisted, and by flying with the captured German equipment, RAF pilots discovered a beam emanating from Cleves and directed on Derby in the East Midlands. This sent a cold chill down the spines of those involved. Derby was the location of the Rolls-Royce factory producing Merlin engines for Spitfires – vital for the Battle of Britain. This was probably the single most important target in Britain. Identification of this beam produced a near panic that the Germans could now bomb at night factories of such major importance to Britain's war effort. Scientists from the Air Ministry and the Telecommunications Research Establishment near Swanage in Dorset worked together at full speed. By the middle of August, all the *Knickebein* transmitting stations across northern France had been identified (the Cleves station had moved to Calais). Then the boffins found ways to jam the

beams electronically, so the bombs would be dropped not on
the intended targets but on open fields some fifteen or twenty
miles away. The counter-measures were known by the code
name 'Aspirin', as they helped clear the headache caused by
Knickebein. By the time the Blitz began in earnest on 7
September with the big raid on the docks and East End of
London, the first phase of the Battle of the Beams had been
won.

Of course, the Germans soon realised that their beams had
been identified and were being distorted. A new chapter in the
scientific war unfolded when the Luftwaffe transferred to a
new system called 'X-Geraet' (or 'X-Apparatus'). This was a
more sophisticated device that used five very high-frequency,
short-wavelength beams to guide the aircraft. When these
beams intersected with cross-beams, a trained navigator could
identify a target with great accuracy – it was estimated down to
about one hundred yards. This system was used by only one
formation, a special 'Pathfinder' group, Kampf Gruppe 100.
The planes of KGr 100 would identify the target and drop
incendiaries and the rest of that night's bombing force would
then aim their explosives at the area of the flames.

On the evening of 5 November 1940, a Heinkel from KGr 100
crashed on the beach at West Bay, near Bridport in Dorset. As
the sea lapped around the bomber a dispute ensued between
the army, who turned up to salvage it, and the navy, who
claimed that as it had landed in the sea the prize was theirs. As
a consequence, the vital electronic equipment was initially lost
and it took several days to find it. Even when it was finally
recovered it proved difficult to piece together how it worked. It
seemed the Germans transmitted eight beams, but only five of
these were used in each raid. And they used two different types
of signal, which the British called 'fine' and 'coarse'. But jamming
or distorting the beams still proved impossible.

Using Ultra decrypts and a close study of how the beams were set up, by mid-November the boffins knew that a big series of raids was coming under the code name 'Moonlight Sonata'. But they were unable to predict the targets and still could not jam the beams. Sure enough, on the night of 14 November, the Luftwaffe mounted its biggest raid yet outside London when it bombed Coventry. The raid was one of the worst of the Blitz. In addition to the hundreds of civilian casualties, the cathedral and more than twenty factories were destroyed. A persistent myth still surounds this raid that Churchill knew the target was Coventry but failed to warn the city or order the jamming of the beams so as not to give away the fact that British boffins understood how the German system worked. It is clear from the evidence of those working long hours to understand and distort the beams that this was not the case. In fact, Churchill thought the raid was heading for London. He ordered his secretaries into underground shelters, telling them they were too young to die. Churchill himself could not sleep and spent much of the night on the Air Ministry roof, looking out for the expected bombers.[20]

By early 1941, a successful means had finally been found to block the X-Geraet system. The stronger counter-measures this time were known by the code name 'Bromide'. The Germans would turn on the beams in the evening before a raid and from this Jones and his team were eventually able to identify what that night's target would be. But the Germans, once again, soon developed yet another device, the Y-Geraet, or Y-Apparatus, an early version of which had been described in the Oslo Report. It used a highly sophisticated form of radar to guide an aircraft to its target. This system was identified and disrupted by the British boffins more quickly. In May 1941, it is thought that the twisting of these signals inadvertently led the Luftwaffe to bomb Dublin, in neutral Ireland, in error. The see-saw war of one side gaining an advantage, followed by the implementation of counter-measures to block this, continued until that May, when the Blitz on Britain was lifted as the majority of the Luftwaffe were sent east to a new target, the Soviet Union. Churchill estimated that the combination of British counter-measures and simple Luftwaffe inaccuracy meant that 80 per cent of German bombs missed their targets. This was, as he said, 'the equivalent of a considerable victory'.[21] Of course, this meant one in five bombs still hit home, and they caused dreadful losses throughout the winter and early spring of 1940–1. But without the boffins' work, the damage would have been much worse. The Battle of the Beams had proved to anyone who cared to doubt it that this would be a scientific war. But Churchill had already taken a major step to ensure that Britain would be on the winning side.

One of the unfortunate consequences of the big meeting held on 21 June, when Jones had first related to Churchill and the RAF bosses the story of the German *Knickebein* beams, was the resignation of Sir Henry Tizard. With Lindemann's close association to Churchill, their old rivalry re-emerged after he became Prime Minister and Tizard realised his position had become untenable. He also accepted that in not backing Jones in his suspicion that the Germans were using beams he had been wrong. He withdrew from his position at the apex of many government defence committees. But Churchill, unlike Lindemann, was not a man to hold a grudge. He recognised Tizard's ability and soon came up with a new mission for him.

During the summer of 1940, Churchill worked hard to persuade Roosevelt to increase the US commitment to the British war effort. We have seen that he was partially successful in this but at the end of July, Churchill personally took a dramatic and courageous decision that would have long-term consequences for Anglo-American cooperation during the war and for long after. With the threat of invasion and the possibility of defeat still real, Churchill decided to share Britain's scientific secrets with the Americans and the Canadians. Behind this extraordinary decision lay the idea that if Britain were defeated, then at least the New World could continue the fight using the latest technological advances. It was not an altruistic gesture. It was motivated by the desire to bring America further into the war. But for a nation at war to supply its top secrets to a non-belligerent country was an act of faith unique in world history. Churchill asked Tizard to chair the vital mission to the United States that would offer up those secrets.

Tizard was delighted with his new task. He hoped that he would succeed in bringing 'American scientists into the war before their Government' and that he could encourage the Americans to reciprocate by sharing some of their advances with Britain.[22] He gathered together a series of papers representing the most advanced thinking of British scientists on miniaturised radar which was being developed for air-to-air use, chemical warfare and on explosives. In addition, he boxed up and took with him actual examples of a variety of new devices. These included three power-driven gun turrets that were later used in bombers; proximity fuses that ignited when they came close to an aircraft, which the United States were to build in huge numbers; and a new predictor for the multiple-firing Bofors gun. But most important of all was a small black box containing the jewel in the crown of British science in the summer of 1940, the cavity magnetron.

The cavity magnetron had been developed by two scientists at Birmingham University, Professor John Randall and Dr Harry Boot, earlier in 1940. Radar, despite its great value, was limited by the inability to send out really short-wavelength radio signals. The cavity magnetron was a valve that overcame this problem by turning high voltages into short-wave signals of immense power. When combined with a receiving valve, the cavity magnetron revolutionised the power and range of a radar set almost overnight. Radar could now pick up and identify the movement of people, the positions of cliffs to assist in the radar navigation of naval vessels, and even the location of the conning tower of a submarine. And the way was opened for its use within aircraft and ships for blind navigation during fog or darkness. And radar's range was extended from forty or fifty to well over a hundred miles. The cavity magnetron was, literally, a war-winning device. When the box containing it was opened and it was shown off to a group of American scientists in a Washington hotel, they were almost blown away by what they saw. The official historian of the American scientific war later described this as nothing less than 'the most valuable cargo ever brought to our shores'.[23]

Churchill's decision to share Britain's secrets with America had long-term consequences of immense importance. A Scientific Office was opened in Washington to coordinate the further exchange of scientific ideas. Delegations of senior American officers and top scientists travelled to Britain to discuss ways of collaborating should the United States enter the war. America not only became a closer ally (Churchill's primary objective) but could now put its vast industrial muscle behind the development and manufacture of some of the latest scientific devices on a scale way beyond anything that was possible in Britain for the rest of the war.

In the summer of 1940, after France had fallen and with the Battle of Britain raging, it became obvious that heavy air raids would soon be made on Britain. But the state of the nation's anti-aircraft defences was poor. There were too few guns, and anyway the process of trying to hit a fast-moving target when it would take many seconds for the ack-ack shell to reach the right altitude, was still very hit and miss (mostly miss). Giant concave sound detectors were used to try to establish the height of enemy aircraft, but these were hopelessly inefficient. The officer in charge of Anti-Aircraft Command, General Frederick Pile, was an imaginative man keen to find any means to improve the accuracy of his gunners. After discussions with Tizard, it was decided to recruit a group of young scientists from the universities and industry from the Central Register drawn up before the war to tackle the problem. They were led by Professor Patrick Blackett, a leading physicist from Cambridge who had been an officer in the First World War and understood the army mentality. The team became known as 'Blackett's Circus'.

The Circus immediately got to work to try to bring scientific principles into the operation of the anti-aircraft batteries. Radar was used to try to predict the path, speed and altitude of enemy aircraft. But unlike the radar used by the RAF in the tall towers of the Chain Home sites, the portable radars used by the army alongside the ack-ack guns were never far off the ground and so were subject to massive interference from nearby buildings and even hills and valleys in the surrounding landscape. Experiments were carried out and it was discovered that placing wire netting around the radars provided a uniform reflecting surface which cut right back on interference. Within weeks, the army had requisitioned nearly all the available stocks of wire netting in the country and this was now laid out around the mobile radars at the batteries.

The second task was to link the guns' firing control with the radar signals to ensure that the batteries could fire at the position the enemy aircraft were going to arrive at in the number of seconds it took for the shells to reach the correct altitude. Masses of statistics were gathered as night after night the men of Blackett's Circus looked for ways to improve on the gun sighting mechanisms. Work was done in Richmond Park to calculate the optimum layout for a battery of guns and the best way to concentrate fire. Throughout the winter of 1940–1, as German bombers flew over almost nightly to blitz the cities of Britain, the men of Blackett's Circus observed the enemy raiders and worked out the best way to hit them.

This process of applying scientific, often mathematical, principles to observe, assess, review and ultimately to improve military operations was called Operational Research. It began to play an important part in the war effort and soon spread from the army to the navy and the RAF. All the services had their traditional ways of doing things, and the boffins of Operational Research were able to observe and assess these and sometimes find more efficient ways of achieving results. In 1941, Blackett summed this up by saying that Operational Research could encourage numerical or scientific thinking on operational matters in order 'to avoid running the war on gusts of emotion'.[24] This was at the heart of the wizard war: that scientific analysis could achieve better results than the time-honoured, traditional way so valued by soldiers, sailors and airmen.

Blackett himself went on to work for the navy and his Operational Research team helped in the deadly Battle of the Atlantic. It was found that aircraft painted white rather than the traditional black were more difficult to see from the sea. After the planes of Coastal Command were repainted, the number of U-boat sinkings increased. It was also discovered that something as simple as resetting depth charges to go off at twenty-five rather than a hundred feet resulted in many more kills. Indeed, losses went up so dramatically that captured U-boat crews said they thought the Royal Navy had developed a new weapon against them. At the height of the Battle of the Atlantic, Churchill became so concerned about the U-boat menace that he held fortnightly meetings at Downing Street which Blackett would attend in order to pass on the latest summaries and observations from his team (see Chapter 7).

The RAF had already embraced a form of Operational Research with the integration of radar technology into Fighter Command. The filter rooms which processed the inflow of information about the approach of enemy bombers and the command centres with their plotting tables and the movement of markers across maps that enabled orders to be sent to scramble the fighter squadrons had all been planned before the war. Soon Bomber Command established its own Operational Research unit, which again came up with a number of suggestions for improved operational efficiency. It was calculated that massed bomber attacks had more impact than a large number of scattered attacks. This thinking led to 'thousand-bomber' raids from May 1942. Scientists analysed how planes were shot down. Research found that most bombers returning from sorties had only one or two shell holes and were not structurally damaged. It was realised that most planes went down because of fires in their fuel tanks or the ignition of petrol vapour generated during a long flight. If an inert gas like nitrogen could be injected into the fuel tanks, then they would be less likely to burst into flames if hit by shrapnel from anti-aircraft fire. This discovery had long-term consequences, but only a limited improvement was possible during the war.[25]

Operational Research units were created in RAF commands overseas. One important unit was set up in Cairo to advise on issues relating to the war in North Africa and over the Mediterranean. But there was a shortage of scientists, particularly mathematicians, so many young men straight from university were put into uniform and rushed out to the Middle East.[26] From here, important work was done on improving Allied operations in the Mediterranean by assessing the strike rate of bombs versus torpedoes against enemy ships (torpedoes had a higher success rate), and discovering the best way to find and sink U-boats. Later, this group learned that orange life-vests and life-rafts were easier to spot in the sea than the traditional yellow ones. This apparently simple discovery would save the lives of hundreds of downed airmen long after the war was over.

The Operational Research teams were only ever advisers. They were often placed in command centres and worked along-side staff officers. Their research sometimes confirmed that the traditional ways of doing things were the best. But when they did come up with new recommendations it was still up to the generals, admirals and air chiefs to decide whether or not to implement them. The fact that the military commanders often did follow the scientists' advice is a sign of the changing balance between science and the military. The wizards were having a bigger impact than anyone could have imagined at the start of the war.

Another key player in Operational Research was Professor Solly Zuckerman. He had started his career as a zoologist and had done important research into the lives of apes and monkeys. When the war began, he was studying anatomy at Oxford. He and some colleagues wrote a short book, *Science at War*, in two weeks at the start of the conflict. Rushed out as a Penguin Special, this was almost a manifesto for the new science of Operational Research. Zuckerman went on to assess the effects of bomb blast on the human body and on buildings during the Blitz. He found that the body was far more able to withstand the blast than had previously been imagined and that most casualties were caused by indirect effects, such as the collapse of buildings. He began to calculate the number of casu-alties likely to be caused by the dropping of particular weights of bombs. Then he went to North Africa to survey bomb damage caused by air attacks on towns captured from the Germans and on military convoys crossing the desert. He used captured German documents as well as ground observations to make his assessments. Later, he looked at the impact of bombing on rail communications in Sicily and mainland Italy. His findings would play an important part in the debate about the bombing offensive which is dealt with in Chapter 8.[27]

Of course, all of this work went on without the direct involvement of Churchill. But his support for men of science, his enthusiasm for change and new thinking, and his desire for the military to keep up with the work of the wizards, created the climate in which this particular seed could grow into a forest. This all stemmed from his War Lab. However, Churchill's loyalty to the most prominent member of that group was both a hindrance as well as a help to the war effort.

In 1942, Churchill offered Lindemann a peerage and he became Lord Cherwell (the name by which he will be referred to from now on). Churchill also formalised his position in government by making him Postmaster General. But Churchill had become over-reliant upon Cherwell's scientific advice. It's unfortunate that Cherwell's abrasive personality meant he fell out with so many distinguished men. And of all those who Cherwell vehemently opposed, Sir Henry Tizard was without doubt the greatest loss to the war effort. Churchill was usually a good judge of character, and he should have insisted that Tizard remain in the positions he had occupied with real distinction since before the beginning of the war. However, the Prime Minister's friendship with Cherwell blinded him to Tizard's qualities and he allowed Tizard to take a back seat after 1940. Without doubt this was a great loss to the British war effort. Tizard graciously summed up the position of the new science when he asked a parliamentary committee in February 1942, 'What Prime Minister of England ever had a scientific adviser continually at his elbow?' And he went on to observe of wartime Britain: 'there is hardly a phase of the national life with which scientists are not associated. In fact a fighting friend of mine said that he could hardly walk in any direction in this war without tumbling over a scientist who had got in his way.'[28]

But by this time, Tizard was under-utilised by the War Lab. He held a variety of minor roles, but none of them compared with his earlier work on air defence or his mission to the United States in the summer of 1940. This had been one of the most important scientific journeys of the war. But there was yet another secret that Britain still had to share with America, one which would have even greater long-term consequences.

Before the First World War, Albert Einstein had reasoned that atoms, the basic unit of matter, were held together by forces that if released could produce huge amounts of energy. In the late 1930s, German scientists in Berlin had succeeded in splitting the atom. Then, in early 1940, two German émigrés working in Britain on ideas about nuclear fission and how to release the power of the atom, Rudolf Peierls and Otto Frisch, made an extraordinary claim. They calculated that the amount of uranium needed in a bomb to unleash the energy equivalent to about a thousand tons of high explosives could be measured not in tons, nor in hundreds of kilograms, but was as small as one pound, less than half a kilogram. Their findings were passed on to Tizard, who at that point was still chairing the Committee for the Scientific Study of Air Warfare. Tizard discussed the possibility of finding some form of military use for this remarkable equation with Professor George Thomson in his rooms at Oxford. Thomson led a group of scientists nicknamed the 'Balliol Beagles' – so called because they always seemed to be chasing after the latest scientific ideas. Tizard decided to form a committee under Professor Thomson to examine the possibility of producing an atomic bomb. For the first time anywhere in the world, a government committee now began to consider the possibility of producing nuclear weapons.

Soon after Thomson and his small team of experts began their deliberations, a strange message arrived from an exiled German physicist working in Sweden. It read: 'Met Niels and Margherita recently. Both well but unhappy about events. Please inform Cockcroft and Maud Ray Kent.' Niels was Niels Bohr, the distinguished Danish physicist who had worked on atomic science before the war. Cockcroft was Sir John Cockcroft, a brilliant Cambridge scientist and Nobel prize-winner who was already a member of Thomson's committee. But no one knew who Maud Ray Kent was. It was decided that the message was probably cryptic and could possibly be an anagram for 'Make Ur Day Nt' – a clue that the Germans were developing their own atomic bomb and that the British needed to speed up. Thomson's committee was renamed the Maud Committee and increased the pace of its work through the summer, autumn and winter of 1940–1.

In July 1941, the committee submitted its report, concluding that it was possible that an effective atomic bomb could be produced within two years. When Cherwell read the report, he was sceptical as the science was still unproven and the production of an atomic bomb would involve a massive reallocation of resources that were more profitably committed to other wartime technologies. However, he concluded to Churchill: 'I am quite clear that we must go forward. It would be unforgivable if we let the Germans develop a process ahead of us by means of which they could defeat us in war.'[29] Churchill agreed with the need to keep ahead of anything the Germans might be developing, so he set up a top-secret organisation under the code name 'Tube Alloys' to develop a nuclear bomb.

However, far more important than the creation of the British project was Churchill's agreement to forward the Maud Report to the Americans. The report was filed away in the United

States for several months until James B. Conant, the president
of Harvard University and a senior scientific adviser to the US
government, got to read it. Finally, in October 1941, he and
other scientists persuaded President Roosevelt to commit the
US government to developing and building an atomic bomb.
Roosevelt offered to include the British in this effort but
Churchill refused, believing that his scientists were ahead of
their American counterparts (at this point, they were).
However, once the United States came into the war after Pearl
Harbor in December 1941, the Americans' work on their atom
bomb went up a gear and rapidly pulled ahead of anything
going on in Britain. In September 1942, General Leslie Groves
was put in charge of the project, which now came under the
supervision of the US Army. Three months later, the first
nuclear reactor was built in Chicago. But the work involved in
making a bomb proved far more complex than anything imagined by Thomson and his team in Britain. The Manhattan
Project, as the US research came to be known, developed into
one of the biggest scientific operations of all time, employing
120,000 people across 37 different research sites. Robert
Oppenheimer led the scientific work and finally, four years
after the Maud Report had been produced, in July 1945, an
experimental bomb was successfully tested in Alamogordo,
New Mexico. One of the scientists present at this test, code-
named 'Trinity', wrote:

> Suddenly there was an enormous flash of light, the
> brightest light that I have ever seen or that I think anyone
> has ever seen . . . as we looked toward the place where the
> bomb had been, there was an enormous ball of fire which
> grew and grew, and it rolled as it grew . . . A new thing
> had just been born; a new control; a new understanding of
> man, which man had acquired over nature.[30]

The war against Germany had already been won two months earlier. The war against Japan would be over within a month.

The atomic bomb had a strange lineage: from Germany to Britain, then through Tizard, Thomson and Cherwell to Churchill, and from him to the United States and into the full Manhattan Project. Only America had the vast resources needed for the development of a weapon that would transform world politics and the balance of power for the next fifty years. But Britain's War Lab had played a key role in starting the ball rolling. For good or ill, the partnership between scientists and soldiers had helped to produce the most destructive weapon known to mankind.

6

The Generals

Winston Churchill was a hard taskmaster, particularly for the military chiefs with whom he worked closely. He was demanding, demonstrative, convinced that he was always right, and kept them up half the night. He constantly felt that his generals, admirals and air marshals did not show enough aggressive instinct and that it was him against all of them. Overall, most of them admired him but felt they had the specific knowledge about the situation in the field that meant they were better judges than he was of what was possible and what was impossible. At best, Churchill viewed this as obstructive; at worst, plain defeatist. In one angry outburst in October 1941, he shouted at them: 'I sometimes think some of my generals don't want to fight the Germans!'[1] The Chiefs of Staff were all central players in the War Lab. But did Churchill cajole and bully them too much? Did he get the best out of them or did he push them too far? Building the right relationship with his military chiefs would be essential to winning the war. But it certainly would not be easy for Churchill, nor for his chiefs.

Churchill's military options were limited during the Battle of Britain and the Blitz. A stream of minutes still poured forth from his desk, demanding information, suggesting priorities, proposing objectives and calling for innovations. Many of them, as usual, required 'Action This Day'. But Britain was in a defensive mode and, as Churchill had told the Soviet Ambassador, his principal strategy was just to survive. Nevertheless, his instincts were all for taking the initiative and trying to throw the enemy on the back foot. In July 1940, when the Battle of Britain was about to begin, Churchill ordered the establishment of a top-secret special unit that would encourage sabotage and subversion behind enemy lines, in the hope of generating uprisings against Nazi occupation. This unit was called the Special Operations Executive (SOE) and Churchill famously instructed the minister he put in charge, Hugh Dalton, to 'set Europe ablaze'.

It was an ambitious, aggressive objective but there was little SOE could do to set fire to anything much in 1940. It had limited resources and was hemmed in by sceptical military bosses. But Dalton understood his mission. He said: 'Regular soldiers are not men to stir up revolution.' The idea was to drop small groups of well-trained saboteurs and assassins into occupied Europe. These would then act in tiny cells to hit at key targets and create a level of mayhem out of all proportion to their numbers. Churchill had long been keen on guerrilla tactics since he had observed them first hand as a young officer in Cuba and South Africa, where the Boers had used these tactics with great effect. With his lifelong enthusiasm for unorthodox methods of warfare, it is not surprising that he set up something like SOE so early in his premiership. He called it the 'Ministry for Ungentlemanly Warfare' and it was all very exciting cloak-and-dagger stuff.[2] But did it in reality do any good?

The people of occupied Europe were soon to feel the full force of the Nazi jackboot. Any uprising was met with ruthless and vicious reprisals. Dozens of innocent men and women would be killed for every action taken against the occupying forces. Thousands of others would be sent to concentration camps. Churchill wanted to provide a beacon to inspire and motivate uprisings while providing essential logistical support. The first parachute drops took place into occupied Poland in 1941. Later that year came operations in Norway and Sweden, but it was only in Yugoslavia that SOE could claim a significant impact. Even then, the eminent military historian John Keegan has claimed that SOE was an expensive and misguided failure, costing far more in the loss of innocent lives than it achieved in positive results.[3]

While Britain's back was very firmly against the wall, Churchill created two other new military forces with the intention of striking soon against Nazi-occupied Europe. The commandos (the term was first used by the Boers for their forces that struck behind enemy lines) were put under the command of Churchill's old hero from the First World War, Sir Roger Keyes. The first commando raids were combined operations with SOE against Norway to capture key pieces of Enigma technology in 1941. And, against much opposition, Churchill also insisted on the formation of a new paratroop regiment. The army chiefs argued that with the invasion scare at its height, skilled combat soldiers could not be spared for new units. But Churchill insisted. A call for volunteers went out and several hundred men came forward. Within a year, the core of the 1st Airborne Division had been trained up. Churchill kept up to date with developments in their training and watched a parachute test drop in person in July 1941. SOE, the commandos and the airborne units would all have notable successes later in the war.

With Cherwell's support Churchill also encouraged the development of a small experimental group to explore new forms of explosives. This was led by two maverick inventors, Major Millis Jefferis and Stuart Macrae, the ex-editor of *Armchair Science* magazine. Both men were brilliant and radical thinkers who over the next few years came up with a run of remarkable and slightly wacky devices, ranging from tiny booby traps to heavy guns, most of which had improbable names like the Kangaroo Bomb and the Beehive. They devised a magnetic naval limpet mine that was used in several commando raids, a sticky bomb that would attach to armour for five seconds before exploding, and a form of mortar that could fire a ring of bombs in a circular pattern against a U-boat, known as Hedgehog. By the end of the war, thirty-seven U-boats were confirmed as having been sunk with Hedgehog. Jefferis also invented a shoulder-fired anti-tank gun that eventually went into production as the PIAT gun and became the army's most effective infantry-operated anti-tank weapon. This was just the sort of unconventional, out-of-the-box thinking that Churchill loved to encourage in his War Lab.

Sometimes the work they did went wrong, and on at least one occasion there was an explosion that destroyed part of the unit's armoury at Whitchurch, near Aylesbury. Needless to say, the War Office and the Ministry of Supply were opposed to such unorthodox experiments and on several occasions tried to close down the whole operation. At one point, Churchill put the group under Ismay's authority and gave him instructions to keep an eye on it. Later, it came under Cherwell's direct control. It became known as 'Churchill's Toy Shop', and the Prime Minister loved visiting Whitchurch to see the latest inventions. On one occasion, Churchill was described as being 'like a small boy on holiday'.[4]

In the summer of 1940, the Mediterranean theatre, including

North Africa, was the only location where there was a remote possibility of taking the sort of aggressive action that Churchill longed for. And so campaigning here grew to obsessive proportions in Churchill's mind. There was a strategic debate about the Mediterranean. The First Sea Lord, Admiral Dudley Pound, proposed abandoning the eastern Med and the long-held British naval base in Alexandria to concentrate on Gibraltar. In July, Churchill vetoed this and decided to reinforce the garrison in Egypt against a likely attack by Italy. Even in this darkest hour, as the battered remnants of the British Army, hauled from the jaws of catastrophe at Dunkirk, desperately reassembled to prepare to defend the country from possible invasion, Churchill persuaded his Chiefs of Staff and the War Cabinet to send men and *matériel* to Egypt. About half of the best armour in the country, over 150 tanks, along with anti-tank weapons, anti-aircraft guns, field artillery, rifles and ammunition were loaded up. It was a courageous, possibly foolhardy, decision. It is probably a sign that Churchill never really believed that a German invasion was likely. The Ultra decrypts that he read often encouraged him in this view. But it was certainly going against the grain to deplete the national defences at such a critical time. Churchill later wrote that the decision 'was at once awful and right', but once he had convinced the other military and civilian chiefs, 'No one faltered.'[5] Churchill wanted to get the supplies to Egypt as quickly as possible and proposed sending them through the Mediterranean. Here he was overruled by the service chiefs, who argued that the route posed too great a risk to such a valuable cargo. After a fight, Churchill demurred and the convoy went by the long route around the Cape of Good Hope. It did not arrive in Egypt until the end of September.

As we have seen, the Commander-in-Chief in North Africa General Wavell came back to London in August 1940 to discuss strategy. Churchill was not impressed with him and his taciturn manner and was tempted to replace him. But he had not done so. Wavell returned to his command in Cairo, and he was there to receive the reinforcements when they arrived in September. Churchill then, once again, started to press for immediate action. Wavell insisted on delaying until everything was ready. In early November, reconnaissance aircraft photographed the assembly of the Italian Mediterranean Fleet in the harbour at Taranto. On the 11th, aircraft from the carrier HMS *Illustrious* attacked the Italian fleet with torpedoes. Three battleships and one cruiser were hit. The resounding success of this attack helped shift the balance of naval power in the Mediterranean in favour of the Royal Navy. This surprise attack upon the Italian fleet in harbour was also the inspiration for the Japanese admirals a year later when planning their assault upon the American Pacific Fleet at anchor in Pearl Harbor.

During the autumn of 1940, Mussolini invaded Greece and an Italian army launched an attack on Egypt. The invasion of Greece was stalled by the valiant defence of the Greek Army. Then, on 9 December 1940, Churchill at last got his long-awaited land offensive, by what he called the 'Army of the Nile'. General Richard O'Connor led this modest affair, known as 'Operation Compass', with only two Allied divisions. Nevertheless, the offensive was a stunning success. By the time it came to a halt in February 1941, O'Connor had advanced five hundred miles along the North African coast through Cyrenaica in Italian Libya as far west as Beda Fomm. The assault had routed ten Italian divisions, and had captured 125,000 prisoners along with 400 tanks and 1200 guns. The newsreels contained footage of endless lines of Italian prisoners as far as the eye could see. Not surprisingly, Churchill eagerly embraced the good news, which was pretty well the first British triumph of arms on land in the war so far. He wrote to President Roosevelt, and to the Commonwealth prime ministers, as he still needed the support of their armies in the Middle East. But this was just the beginning of the see-saw war in North Africa, which would be the principal theatre of operations for the British Army for the next two years.

It will be remembered that Churchill had done much to foster Roosevelt's support during the dark days of 1940. This had led to America's supply of the second-hand destroyers Churchill was so desperate for, along with tanks, artillery and other weaponry, much of which had gone to buttress the army in Egypt. However, before committing himself to more aid, Roosevelt wanted to gauge the British war effort and particularly the determination of the British Prime Minister. In January 1941, he sent his close friend Harry Hopkins as his personal emissary to Britain. Churchill really put himself out to charm Hopkins and spent twelve evenings with him. His schedule was carefully stage managed by Brendan Bracken. There was a visit to Dover to review the defences and Hopkins peered across the narrow Channel to occupied Europe; he toured Britain and saw stout-hearted defenders everywhere, alongside the damage caused by the Blitz; and he even travelled far north to Scapa Flow to inspect the mighty fleet gathered there. He also met the War Cabinet and the Chiefs of Staff.[6] Churchill did not share with him the secret of Ultra. During his visit, further intercepts came through, suggesting that the German invasion had been called off. But Churchill kept these to himself and instead talked up the threat of invasion, which he knew would have the biggest impact upon the Americans.[7]

Hopkins was enraptured by Churchill and entirely won over. At the end of his visit, he reported back to the President:

I have got a reasonably clear perception not only of the physical defences of Britain, but of the opinions of the men who are directing the forces of this nation. Your 'former

Navy person' is not only the Prime Minister, he is the
directing force behind the strategy and conduct of the war
in all its essentials . . . The spirit of this people and their
determination to resist invasion is beyond praise. No
matter how fierce the attack may be you can be sure they
will resist it, and effectively.[8]

With the Blitz still on and with no end in sight, this was a fascinating verdict on Churchill's Britain in early 1941.

Roosevelt, reassured, pressed ahead with his support. The
Lend-Lease Act passed through Congress. America could now
build whatever was needed and lease it to Britain, who would
pay up in full after the war was over. Before this, Britain must
pay all the debts it could in gold and sell its commercial assets
in the United States. It was a tough deal that totally drained the
nation's reserves and in one sense left Britain bankrupt. But it
marked a long-term commitment by the United States to
Britain's war effort. It enabled Churchill to keep fighting the
war. In a broadcast to America he used the famous phrase:
'Give us the tools and we will finish the job.' In the House
of Commons he claimed that Lend-Lease was a 'monument of
generous and far-seeing statesmanship'. And in a telegram
to Roosevelt he wrote: 'Our blessings from the whole British
Empire go out to you and the American nation for this very
present help in time of trouble.' In private, however, he admitted that the sale of Britain's assets meant that 'we are not only
to be skinned but flayed to the bone'![9]

The collapse of Mussolini's armies in both Greece and Libya
now drew Hitler into the Mediterranean. In a sense this was a
victory for Churchill's Mediterranean policy. At least he had
found somewhere for offensive action to create an impact. But at
the time, he once again felt that he was being let down by over-
cautious thinking among his army chiefs. Having advanced far

into Libya, O'Connor's offensive was called off because the intelligence coming through Ultra revealed that the Germans were preparing to intervene in Greece. At a meeting in mid-February, Churchill lost his temper with the CIGS, General John Dill, who told him troops could not be spared for Greece. Churchill could not fathom why, after orchestrating this great victory, with 300,000 men and all the supplies he had been sent, Wavell was unable to defend Greece and maintain the offensive campaign in North Africa at the same time. 'What you need out there is a Court Martial and a firing squad!' he shouted. Dill remained tight-lipped under the pressure of this Churchillian tantrum, but later he wished that he had replied: 'Whom do you want to shoot exactly?' But he didn't think of this until afterwards.[10]

The British intervention in Greece was a sorry affair. Churchill and the Chiefs of Staff ordered Wavell to halt the offensive in Libya and to dispatch troops to Greece. Reluctantly he did so. In April 1941, when the powerful German assault began, the British, Australian and New Zealand troops were both vastly outnumbered and outfought. They were forced back and on 17 April, Churchill sent agreement for their withdrawal in a sort of Greek Dunkirk. About 50,000 of the 62,000 men sent to Greece were evacuated by the Royal Navy, but they left behind most of their guns, tanks and transport. Crete was now reinforced and detailed plans were picked up in Ultra for a German airborne invasion of the island. However, the army and the intelligence people back in London, particularly Stewart Menzies, 'C', were desperate not to give away the fact that they had deciphered the German military messages. So the defence of Crete was not reorganised in a way that would have best prepared it for the German assault. General Bernard Freyburg, in command on Crete, was given full access to the Ultra decrypts and details of the planned airborne invasion, but effectively he was told not to act on them. 'The authorities in England would prefer to lose Crete rather than risk jeopardising Ultra,' Freyburg said later.[11] The New Zealand defenders fought bravely, but in just a few days at the end of May they suffered fifteen thousand men killed, captured or wounded, and Crete was duly lost. For the German invaders, however, success came at a high price. Seven thousand were killed during the assault, and German airborne troops were never again used to assault defended positions.

It is sometimes claimed that these actions in Southern Europe at least delayed Hitler's assault on Russia, with the key consequence that his armies were unable to capture Moscow before the onset of winter. It's unlikely in retrospect that they did have this effect. More effective was an SOE-inspired anti-German coup in Yugoslavia. This brought Hitler into the Balkans and prompted his assault on Yugoslavia in order to secure his southern flank. It resulted in years of terror and repression in the Balkans. And it was this action that delayed the start of his assault upon the Soviet Union by a crucial five weeks.

Another consequence of Hitler's attempt to prop up his ally Mussolini was the despatch to North Africa of General Erwin Rommel, the man Hitler described as the most daring commander in his army. Although initially Rommel had only minimal forces at his disposal, these forerunners of the Afrika Korps soon seized the initiative and rolled back the great British advance across Libya. By the spring, Rommel and his combined German and Italian troops had pushed the British Army right back to the Egyptian border. To add to the humiliation, General O'Connor was captured as the British front collapsed. This presented Churchill with yet another failure of British arms. Only the port city of Tobruk managed to hold out against Rommel's lightning advance, and was left behind as a a besieged enclave.

Rommel sent messages to Berlin that his troops were exhausted and needed to rest. Two days later, the Ultra decrypts of these messages were in the faded yellow case on Churchill's desk. Once again, the Prime Minister chafed at the bit and demanded that Wavell launch a fresh assault. Another convoy of supplies, including more than two hundred new tanks, were sent to reinforce the Army of the Nile. This time, though, on Churchill's insistence, the convoy successfully took the quick route across the Mediterranean. Churchill signalled to Wavell: 'I have been working hard for you in the last few days ... no Germans should remain in Cyrenaica by the end of June.'[12]

On 15 June, Wavell's new offensive, Operation Battleaxe, was launched. But Rommel had managed to decipher British Army messages sent from Middle East Command. Forewarned, he placed his front-line troops on full alert. They were waiting and ready for Battleaxe. After three days of fighting, Rommel cleverly outmanoeuvred the British and Wavell called off the failed offensive.

Churchill was furious. He felt like he was carrying the whole war effort on his shoulders. It seemed to him that he was facing cautious Chiefs of Staff, admirals who complained about exhausted ships, and generals in the field who lacked the will to fight. He now came to think he should take military command himself. On the other hand, for his generals the constant interference and sending of detailed memos with minute instructions about operational matters was severely wearing them down. Wavell complained to friends about the relentless 'barracking' from the Prime Minister. The reality was that Churchill did not have the right generals around him. So he did not appreciate their qualities and they were unable or unwilling to do his bidding. Churchill responded by dismissing Wavell immediately and swapping him with General Claude Auchinleck, who was commander-in-chief in India. Wavell was quickly packed off to Delhi, as Churchill put it, to sit 'under a

pagoda tree'. Ismay, an old friend of Auchinleck, warned the
new North African commander about Churchill's ways. His
advice is fascinating. He told Auchinleck:

> The idea that he was rude, arrogant and self-seeking
> was entirely wrong . . . He was certainly frank in speech
> and writing but he expected others to be equally frank with
> him . . . He venerated tradition but ridiculed convention . . .
> His knowledge of military history was encyclopaedic and
> his grasp of the broad sweep of strategy unrivalled. At
> the same time he did not fully realise the extent to which
> mechanisation had complicated administrative arrange-
> ments . . . 'When I was a soldier' he would say, 'infantry
> used to walk and cavalry used to ride. But now the infantry
> require motor cars and even the tanks have to have horse
> boxes to take them into battle.'

Ismay also warned his friend that he would be bombarded by
telegrams on every topic, 'many of which might seem irrelevant
and superfluous'. But he was not to be irritated. Churchill was
carrying the burden of fighting the whole war on every front,
at land, at sea and in the air.[13] Auchinleck was stepping into the
most difficult job in the British Army. Time would tell if Ismay's
advice would bring him more success than his predecessor.

Within days of the failure of Operation Battleaxe, the whole
shape of the war changed dramatically. At dawn on 22 June
1941, Hitler launched Operation Barbarossa, his invasion of the
Soviet Union. For months intelligence reports had predicted the
invasion, so Churchill had had plenty of time to think through
his position. On the evening of Saturday 21 June, he told his
dinner guests at Chequers, including the US Ambassador James
Winant, that he was certain the Germans were about to attack
Russia. He said he would instantly pledge Britain's support to

Stalin. Winant, who had discussed the matter with Roosevelt, agreed that this would also be the US position. John Colville, his private secretary, took a walk with Churchill in the garden after dinner and asked if supporting his old enemy would put him in a difficult position. Churchill replied that 'If Hitler invaded Hell he would at least make a favourable reference to the Devil.'[14]

In the early hours of the following morning, the news came through of the launch of Operation Barbarossa. That Sunday evening, knowing that he had US backing, Churchill announced the position of the British government on the BBC. It was an emotional broadcast, envisioning the noble defence by impoverished Russians of their beloved homeland against 'Hitler's blood-lust'. Churchill declared: 'Any man or state who fights on against Nazidom will have our aid.' He went on, 'The Russian danger is therefore our danger, and the danger of the United States, just as the cause of any Russian fighting for his hearth and home is the cause of free men and free peoples in every quarter of the globe.'[15] Despite a lifetime of personal hostility, and despite all that he had done to undermine the communist state in its early days, the Soviet Union was now Britain's ally – and that was official.

Ultra intelligence had for many months revealed that German troops were massing along the Soviet border, and Churchill had warned Stalin in a personal note of the possibility of invasion.[16] Stalin had received several other warnings of the imminent attack. But in one of the strangest blunders of the war, he ignored all of them and did nothing to prepare his vast land army or his air force. As a consequence, when Barbarossa was launched, the German military thrust forward with devastating speed. Three million men led the assault on the Soviet Union along a border of 450 miles, from the Baltic to the Black Sea. The Germans had a total of 120 divisions in three gigantic army groups supported by 3350 tanks. On the first morning, the Luftwaffe attacked the Soviet Air Force while it was still on the ground, destroying many aircraft before they were even able to get airborne. The Soviet Red Army was huge, but no match for the battle-hardened German infantry and panzer divisions. Hundreds of square miles were captured and occupied. One after another, major Soviet cities fell to the Nazi invaders – Minsk, Smolensk, Odessa. Six hundred thousand Russian soldiers were taken prisoner. Hitler spoke of his 'crusade' against communism. He called the Soviet state a 'rotten structure' that when kicked in would come 'crashing down'. It looked as though he would be proved right.

In London and Washington, the general feeling was that the Soviet Union would be overwhelmed within a few months. Barbarossa therefore injected a new intensity into the Anglo-Amereican relationship. Now it was time for a face-to-face meeting. On 4 August, amid tight security, Churchill along with his Chiefs of Staff boarded the newest battleship in the Royal Navy, HMS *Prince of Wales*, at Scapa Flow. Within minutes, the giant battleship slipped its moorings and was soon heading at full speed across the Atlantic. During the voyage, Churchill took his first break of the war. He loved life on board and found time for watching films (particularly historical movies like *Lady Hamilton*), playing backgammon, and even reading a novel (*Captain Hornblower RN* by C.S. Forester). On the morning of 9 August, the battleship arrived at Placentia Bay off Newfoundland. It had been decided to meet on 'neutral' ground rather than in the United States. President Roosevelt arrived on the American heavy cruiser USS *Augusta* and Churchill, refreshed after the voyage, crossed to the American ship, where he was piped aboard. Within a few hours, observers were reporting that he and Roosevelt, who had been corresponding now for twenty-one months, were getting on famously.

The meeting at Placentia Bay provided a giant photo opportunity. On Sunday 10 August, Roosevelt visited the *Prince of Wales* for a church service held on deck. The prayers were carefully selected, as were the hymns, which included 'Onward Christian Soldiers'. The newsreel cameras rolled as the President and the Prime Minister sitting alongside each other led their men in the singing. 'You would have had to be pretty hard boiled not to be moved by it all – hundreds of men from both fleets all mingled together . . . It seemed a sort of marriage service between the two navies, already in spirit allies,' wrote John Martin, one of Churchill's secretaries, in his diary.[17] The meeting is famous for producing the Atlantic Charter, a joint Anglo-American declaration of principles respecting 'the rights of all peoples to choose the form of government under which they live'. It was a grand proclamation of the moral values of the democracies intended to act as a beacon to those under occupation. Its value was largely symbolic. But important business was also done, in great secrecy, at this momentous meeting.

Roosevelt promised to provide aid to the Soviet Union 'on a gigantic scale', and it was agreed to send an Anglo-American mission to Moscow to discuss the USSR's needs and to propose a summit meeting with Stalin. It was also decided to send a strong note to the Japanese, warning them against further encroachments in the Pacific. Critically, for Churchill, Roosevelt and his naval chiefs agreed to provide escorts for convoys across the Atlantic which, if attacked, would respond by firing on U-boats. And the United States would patrol the Atlantic to the west of Iceland. The two sets of Chiefs of Staff had useful meetings. It was clear there was a gulf between their thinking and that the Americans as yet had no plans in place should they become embroiled in the war. But useful personal contacts were made and the basis for future combined operations was laid.

For Churchill, most important of all was the personal time he spent with Roosevelt, building not just a working, professional relationship but a genuine mutual friendship that would be at the heart of the 'special relationship' for the next four years.

Churchill returned from Newfoundland to news of further German advances in the Soviet Union. The Wehrmacht now broke through into the giant, open prairie fields of the Ukraine. It seemed unstoppable. On 21 September, the Germans captured Kiev along with two-thirds of a million Soviet prisoners. Hitler called it the 'greatest battle in world history'. For Churchill, these were bleak times. He worried that if the Soviet Union were defeated, Hitler's triumphant armies would then turn once more on Britain. Again, he called for offensive action in North Africa, but Auchinleck his new commander told him the army would not be ready until November.

It was essential for Churchill to have the right men around him. Wavell had had to go. And so too would General Dill, whom Churchill thought off as being over-cautious, referring to him as 'Dilly-Dally'. It was a few months before the right moment came but later in the year he was replaced as Chief of the Imperial General Staff and sent off to Washington where he was to do an outstanding job. The man Churchill chose to replace Dill as CIGS, the army chief, was General Sir Alan Brooke (later Field Marshal Viscount Alanbrooke). Brooke was part of an Ulster Protestant family with a long tradition of service in the army. One of Churchill's close friends from his army days in India had been Victor Brooke, Alan's elder brother, who had been killed in 1914. And another brother, Ronald, had ridden with Churchill into Ladysmith to relieve the city during the Boer War. Alan Brooke himself had served with distinction in France in the spring of 1940 and had been put in command of the Home Forces in July 1940, when the possibility of invasion loomed. He had impressed Churchill with the way he had reinvigorated the defences of southern England around a more mobile strategy in order to respond with flexibility if an attack came. With frenetic energy he reorganised units that were still reeling from the loss of France. Churchill liked what he saw and the fact that Brooke had the self-confidence to stand up for himself and argue his case against anyone – including the Prime Minister. Churchill respected this and enjoyed the cut and thrust of debate. 'When I thump the table and push my face towards him what does he do? Thumps the table harder and glares back at me. I know these Brookes – stiff-necked Ulstermen and there's no one worse to deal with than that!' Churchill is reported to have said about Brooke to his deputy.[18] After a couple of late night sessions talking together, Churchill offered Brooke the top position.

So began one of the central partnerships of Churchill's War Lab. It endured for three and half years, to the end of the war. They would meet almost daily, and Brooke became Churchill's principal military adviser. To many outsiders it seemed that their relationship was difficult. They argued fiercely together and the atmosphere often seemed tense. Brooke, like his predecessor, was also infuriated by Churchill's micro-management and by his meddling in minor tactical details, the stuff which Churchill of course loved. It always seemed that Churchill wanted to control every aspect of every battle plan. He interfered, it seemed to Brooke, without a full understanding of the situation in the field. Furthermore, Brooke was pushed to the brink of exhaustion by the hours Churchill kept. The Prime Minister would often call meetings late at night or even in the early hours of the morning and he always expected everyone to be at their best. But he would have enjoyed a leisurely morning, reading papers and dictating from his bed. The others would have worked a full, stressful day structured around a conventional time frame. Churchill never seemed to have consideration for the strain that others around him were also under.

But Brooke turned out to be the perfect foil to Churchill. He stood his ground against the Prime Minister's garrulous verbal assaults, and he never accepted a plan or endorsed a policy that he knew was wrong or would needlessly risk the lives of his men. Brooke could be as tough as nails. His nickname in the Cabinet Office was 'Colonel Shrapnel'. Churchill respected him for this. Although the Prime Minister would often cross-examine and criticise everything put before him, sometimes in an aggressive or bad-tempered way, Brooke knew that the very next day he might hear Churchill eagerly putting forward the same plan as though it were his own. And at heart, Brooke had great admiration for Churchill and for his leadership of Britain. Back in May 1941, before he became CIGS, he had the opportunity to observe Churchill at close quarters and had written: 'He is quite the most wonderful man I have ever met, and is a source of never ending interest studying and getting to realise that occasionally such human beings make their appearance on this earth. Human beings who stand out head and shoulders above all others.'[19] After the two had been at each other's throats in some gruelling encounter over the details of a battle plan Brooke would be heard to murmur, 'That man!' And then with a sigh to continue, 'But *what* would we do without him?'[20]

Over the years, both men were worn down by the strain of working together. Brooke often found Churchill petty, unfair and irrational, and his ideas wildly unrealistic. Totally against regulations, he kept a diary, which he wrote up late at night. For him, it provided relief from the stress of the day to scribble down his thoughts. Looking at the original diaries, it's easy to see how he wrote at speed, without much care for grammar or punctuation, almost in a stream of consciousness, to get his feelings off his chest. In one exhausted late-night entry from September 1944, after another bitter row, he recorded:

> We had another meeting with Winston at 12.00 noon. He was again in a most unpleasant mood. Produced the most ridiculous arguments that operations should be speeded up . . . He knows no details, has only got half the picture in his mind, talks absurdities and makes my blood boil to listen to his nonsense . . . Never have I admired and despised a man simultaneously to the same extent.[21]

But the fact is the tension in this love-hate relationship brought out the best in both men. Brooke was confident enough (just) to put up with Churchill's bullying. Churchill, for his part, respected Brooke as a supremely professional soldier, from a long line of soldiers, who always had the army's best interests at heart. And when faced with determined, well-argued opposition, Churchill never overruled Brooke. Churchill's undisciplined genius was tempered by Brooke's tough professionalism. It would be a war-winning partnership.

The very morning that Brooke began to work as CIGS, Auchinleck finally launched his offensive in the North African desert. Within days his men had advanced fifty miles, and on 29 November they relieved the siege of Tobruk. Churchill was delighted with their progress. Then intercepted Ultra messages enabled the RAF to locate and sink two vessels bringing vital fuel supplies across the Mediterranean to Rommel's forces. Churchill had just two ambitions in the closing months of 1941: to bring the United States into the war before the defeat of Stalin's Russia; and to win a victory in the Middle East. He telegraphed Auchinleck in Cairo: 'as long as you are closely locked with the enemy, the Russians cannot complain about no second front . . . the only thing that matters is to beat the life out of Rommel and Co.'[22]

Further Ultra decrypts reported that the German Army in Russia was running out of steam, and its commanders were concerned by lengthy supply lines and the ferocity of the Russian resistance. Churchill passed on details from several Ultra reports to Stalin, including an outline of the full German battle plan, which made it clear that the main German objective was not the oilfields of the Caucasus but Moscow itself. Without letting the Soviets know the source of the intelligence, Churchill forwarded message after message to Stalin. 'Has Joe [Stalin] seen this?' he would regularly ask Srewart Menzies.[23] The Soviets received high-grade information about German plans that they could not have obtained from any other source.

The first light snows fell in September, but the full force of winter swept across Russia in November. German troops had advanced to within thirty miles of Moscow. Advanced patrols could see the spires of the Kremlin through their binoculars.[24] But as the Soviet defence hardened, so the German forces came to a standstill. Then, on 5 December, Stalin did what the Germans had thought was impossible. Using fresh troops from eastern Asia, fully kitted out and trained for winter war, he launched a counter-attack. The German troops, unprepared for the freezing conditions, without winter clothing or suitable weapons, were thrown back. Moscow was saved. And for this winter at least, so was Stalin's Russia.

The weekend after the counter-attack, Churchill was at Chequers. Now his big worry was that Japan was showing signs of aggression towards British and Dutch territories as well as the independent Siam (now Thailand) in South-East Asia. Churchill got in touch with Roosevelt and they agreed to make it clear that an attack upon Siam would be regarded as an attack upon themselves. On Sunday 7 December, Churchill had just finished dinner with the US Ambassador and Averell Harriman, another American diplomat, when they turned on a small radio to hear the BBC's nine o'clock news. Churchill himself almost missed the piece that mentioned an attack by the Japanese on American shipping in Hawaii. The butler came in to confirm that he had heard the report also. Churchill immediately called the White House, and after a few minutes Roosevelt came on the line. 'It's quite true,' the President replied. 'They have attacked us at Pearl Harbor. We're all in the same boat now.'[25]

At this point, the full scale of the daring and unexpected Japanese attack on the US Navy's giant Pacific Fleet as it lay in harbour was unknown. Later, it became clear that four battleships and over two thousand American sailors had been lost that morning. Moreover, Japanese troops had launched simultaneous attacks upon British Malaya, the Dutch East Indies and Siam. The whole world was now at war. But Churchill instantly recognised this as the best news possible, America would at last be Britain's ally. He later wrote that he realised then

> We had won the war . . . Once again in our long Island history we should emerge, however mauled or mutilated, safe and victorious. We should not be wiped out. Our history would not come to an end. We might not even have to die as individuals. Hitler's fate was sealed. Mussolini's fate was sealed. And for the Japanese, they would be ground to powder.

With his staff buzzing around him, Churchill retired to bed, where he 'slept the sleep of the saved and thankful'.[26]

The following day, Britain declared war on Japan. And then Congress declared war. Roosevelt described 7 December 1941 as a 'date which will live in infamy'. Churchill immediately made plans to visit Washington, concerned that vital US supplies might now be cancelled in order to pursue the war with Japan. Then, on 11 December, Hitler chose to declare war on the United States. It was an extraordinary decision. Now it was imperative that the two leaders of Britain and the USA, at long last Allies in the war, should get together. On 13 December, Churchill set sail once again for the United States. It would be a long trip.

During the Atlantic crossing, HMS *Duke of York* was buffeted by heavy storms and made slow progress. En route Churchill received good news of further Soviet counter-attacks in the north at Leningrad and in the south along the Sea of Azov, and bad news of the Japanese attack upon Hong Kong. The long journey and forced inactivity on board gave Churchill and his Chiefs of Staff (the newly appointed Brooke as CIGS was not on this trip) plenty of time to reflect on the global conflict now raging. Churchill wrote three expansive papers on grand strategy. They show remarkable insight into the course the war would actually take. The first focused on the war in North Africa in 1942, arguing that it was essential for the United States and Britain to take control of the Mediterranean. The second was on the war in the Pacific against Japan and correctly foresaw that this would be a maritime struggle followed by a series of island invasions to recapture lost territory. The third addressed the strategic objectives for 1943 and rightly predicted that the conflict on the Eastern Front with the Soviet Union would become the central land conflict of Hitler's war. Although he correctly predicted several major events over the next few years, Churchill was a little optimistic with the timing of these events: he thought the Allied invasion of occupied Europe would occur in the summer of 1943, rather than June 1944.[27]

After ten days at sea, Churchill finally reached the United States and flew on to Washington, where Roosevelt met him and invited him to stay in the White House. They immediately began talks, while the British Chiefs of Staff met with their US counterparts, quickly establishing the basis of how they would work together from now on, a system of combined operations that lasted to the end of the war. They also agreed, after considerable pushing by the British, on the 'Germany First' principle: that is, American resources would be allocated to the defeat of Germany as the first priority in the global war. In retrospect, this was one of the most important strategic decisions of the war. It could so easily have gone the other way. The Americans were outraged at the unprovoked attack by Japan and were tempted to concentrate on the war in Asia with Europe as a secondary diversion. It was a tribute to Churchill and his team, as well as to the global view of the US leadership, that Germany First became the guiding beacon of the war effort.

Much future planning was debated and agreed during the Washington meetings, and once again Churchill enjoyed several hours each day of premium time with the President. They lunched together daily, usually with Harry Hopkins. Dinner was a more social occasion, with others also present. Churchill wrote that 'The President punctiliously made the preliminary cocktails himself, and I wheeled him in his chair from the drawing room to the lift as a mark of respect.'[28] (Churchill preferred whisky and soda to the President's cocktails. Roosevelt was amazed at but tolerant of the amount of alcohol Churchill consumed. However, Eleanor Roosevelt, the First Lady, regarded the Prime Minister as something of an alcoholic.) The bond of friendship between the two leaders grew stronger in the weeks that Churchill lived as Roosevelt's guest in the White House.

On 26 December, Churchill was invited to address both Houses of Congress. He had spent some time polishing his speech and it went down well. The newsreel cameras were there to record one of Churchill's finest, most assured performances. He began by saying, 'I cannot help reflecting that if my father had been American and my mother British, instead of the other way round, I might have got here on my own. In that case, this would not have been the first time you would have heard my voice.' The laughter at this quip turned into profound applause when, speaking of Japan's outrageous attack upon Pearl Harbor, he asked, 'What sort of people do they think we are?' He spoke about the 'long, hard war' ahead and came to a rousing conclusion: 'It is not given to us to peer into the mysteries of the future. Still, I avow my hope and faith, sure and inviolate, that in the days to come the British and American peoples will for their own safety and for the good of all walk together side by side in majesty, in justice and in peace.' These final words brought the cheering congressmen to their feet.[29]

A few hours later, back in the White House, Churchill went to bed but couldn't sleep. As it was a warm evening he tried to open the window. As he did so, he suddenly felt short of breath. He felt a dull pain over his heart which went down his left arm. He quickly recovered, but the following day he reported the incident with some concern to his doctor, Sir Charles Wilson, who was travelling with him. Wilson instantly recognised that Churchill had suffered a mild heart attack, what he called a 'coronary insufficiency'. While he examined Churchill with his stethoscope, Wilson thought hard about what to do. The textbook recovery period for this at the time was six weeks in bed. But Wilson knew that if he instructed Churchill to stop work at this critical juncture it would soon get out 'that the PM was an invalid' with heart problems. It could be the end of Churchill as an effective war leader. On the other hand, if he ignored it and Churchill had another, potentially fatal, seizure, 'the world would undoubtedly say that I had killed him through not insisting on rest'.

The doctor decided that as the attack had obviously been mild he would risk it. He told Churchill, 'There is nothing serious. You've been overdoing it.' When Churchill protested that there was no way he could take a rest now, Wilson replied, 'Your circulation was a bit sluggish . . . you musn't do more than you can help in the way of exertion for a little while.'[30] Despite his age (he was now sixty-seven), the immense strain he was under and the vast workload, Churchill kept going at full speed. Later in the war, he suffered further illnesses, but he was never incapacitated by another heart attack. Wilson had made the right call.

A couple of days later, Churchill visited Ottawa and addressed the Canadian Parliament. Again the speech was recorded on film, which captures the power of his oratory. He referred at one point to the French claim in June 1940 that, without France, Britain would have its neck wrung like a chicken in three weeks. 'Some chicken!' exclaimed Churchill. As the laughter died down he followed this up with 'Some neck!'

Churchill did take a short break for a few days in Florida, where he enjoyed swimming in the warm sea. Swimming was something that particularly appealed to the boyish side of Churchill. 'Winston basks half-submerged in the waters like a hippopotamus in a swamp,' recorded Wilson in his diary, no doubt relieved that at last a couple of days of relaxation had been squeezed into the schedule.[31] Then it was back to the White House, and agreement on a declaration that Roosevelt called the United Nations Pact. This was a follow-on from the Atlantic Charter signed the previous August, a further declaration of democratic principles. There were meetings of real substance about industrial output and a Combined Chiefs of Staff Committee was formed. Finally, it was decided to launch an invasion of French North Africa later in the year to help secure the Mediterranean. It was during this series of meetings that Sir John Dill started to forge an excellent relationship with General George Marshall, the US Army's sturdy and dependable Chief of Staff. When Churchill and the rest of the British team returned to Britain, Dill was left in Washington as head of the Joint Mission, a critical role in the Anglo-American relationship that he enthusiastically and ably fulfilled until his death in late 1944. By then, the man whom Churchill had nicknamed Dilly-Dally had finally proved his worth.

Churchill had been away for three weeks when it was decided business was done and it was time to go home. First he flew to Bermuda in a big Pan-American Boeing Clipper flying boat, and during the flight he took the controls for about twenty minutes. He had not outgrown his fascination with flying. With so many U-boats in the Atlantic and with the urgent need to get back to London, it was then decided that the best option was to continue the journey to Britain in the flying boat. It was an eighteen-hour flight, and towards the end the Clipper drifted slightly off course and dangerously near to the coast of occupied France. It veered north at the last minute and flew towards Plymouth at an angle that made it look like an approaching enemy bomber. Six Hurricanes were scrambled to intercept the Clipper, but fortunately the error was discovered in time and Churchill and his team landed safely.

The following few months saw disaster follow disaster for Britain. The Japanese made dramatic advances in the Far East in their version of the blitzkrieg. On 15 February 1942, the giant fortress at Singapore surrendered (see Chapter 7). It was a humiliating blow to British power and prestige in the East. Churchill described it as 'the greatest disaster to British arms which our history records'.[32] In North Africa, the gains of Auchinleck's offensive were reversed by Rommel. British troops began to retreat once more, although Tobruk again held out, offering some hope. Churchill worried that the British Army in

the Far East and in North Africa did not seem to have the spirit
to fight. Then the German Navy changed the configuration of
their Enigma machine, which meant it was impossible to decipher their messages for most of the rest of the year. Shipping
losses in the Atlantic rose alarmingly. And the convoys carrying
supplies to Russia took a terrible battering from German aircraft
and battle cruisers operating from northern Norway. As the
strain grew, Churchill became very down. 'Papa is at a very low
ebb . . . worn down by the continuous crushing pressure of
events,' wrote his daughter, Mary, in her diary.[33]

With the worsening global situation, Churchill decided to
call another conference with Roosevelt in mid-June. This time
he flew across the Atlantic in the same flying boat he had used
five months before. He spent a couple of days with Roosevelt
in his house at Hyde Park, overlooking the Hudson River. On
20 June both men travelled to Washington in the presidential
train. They were in a meeting in the White House the following
morning when an aide came in and passed a pink slip of paper
to the President. He read it, said nothing and passed it to
Churchill. It read: 'Tobruk has surrendered with 25,000 men
taken prisoner.' Churchill was thunderstruck by the news. He
had placed so much weight on the North African campaign and
now all of his hopes seemed to collapse. He despaired at the
failure of the British Army to hold out even against inferior
numbers of the enemy, at Singapore and now at Tobruk. He
later wrote of this moment: 'This was one of the heaviest blows
I can recall during the war . . . it had affected the reputation of
the British armies . . . I did not attempt to hide from the
President the shock I had received . . . Defeat is one thing, dis-
grace is another.' Roosevelt asked what he could do to help.
Churchill replied that he needed more tanks. The President
immediately summoned General Marshall to join them and
within hours Marshall had come up with a plan to divert three

hundred of the newest A4 Sherman tanks to North Africa. Marshall also agreed to send a hundred artillery pieces. These were soon on fast transports across the Atlantic. They would make a real difference to the North African campaign later in the year. It was a magnificent sign of Roosevelt's friendship for Churchill and of US support for Britain's fragile war effort.[34]

A few days later, Churchill returned home from Washington to face a vote of censure in the House of Commons. Constant news of defeats had created rumblings of discontent against his leadership, and this was the second parliamentary vote of no-confidence that year. The debate offered a sounding board for those who wanted to air their criticism, but this was not a time for a change of leader. Churchill won the vote by 475 votes to 25. On the same day, Rommel's troops reached a small railway station named El Alamein, only eighty miles from Cairo. Here they stopped, for now.

On the suggestion of the soldier son of a colleague who had just returned from Egypt, Churchill decided to visit the troops in the desert to assess for himself the state of the Army of the Nile. After another gruelling flight, this time in an unpressurised, unheated Liberator bomber (Churchill had his oxygen mask specially adapted so he could smoke a cigar while wearing it!), the Prime Minister landed in Cairo. Once again Charles Wilson accompanied him, and he recorded that Churchill arrived 'in great heart . . . A great feeling of elation stokes the marvellous machine, which seems quite impervious to fatigue.'[35]

Churchill soon picked up on the low morale in the 8th Army, and he and Brooke, who also accompanied him on this trip, realised that there must be something wrong with the command of the desert army. After a few days, Churchill decided to replace Auchinleck. He appointed General Harold Alexander as Commander-in-Chief. Alexander was a classic senior officer of the old school, an aristocrat and an ex-Guardsman, not too clever, but always willing. He had a distinguished war record from the Great War and had seemed to sail through life, effortlessly achieving whatever he set out to accomplish, always immaculately dressed and with film-star good looks. He had done well commanding an army corps guarding the evacuation at Dunkirk and was known as a safe pair of highly capable hands. Churchill admired him greatly and was sure he had the diplomatic skills necessary for senior command.

Brooke wanted to appoint General Bernard Montgomery as commander of the 8th Army. But Churchill had reservations about Montgomery and preferred General William Gott, a corps commander who was already in the desert. Brooke regarded Gott as too tired to take on this responsibility and thought Montgomery was more energetic and self-confident. Unusually for Brooke, he gave in to Churchill and agreed to appoint Gott. Then, a couple of days after his appointment, Gott was killed when his plane was shot down by German fighters outside Cairo. Churchill then agreed with Brooke's recommendation and decided on Montgomery to take command of the 8th Army. It proved to be an inspired choice.

Montgomery (better known as 'Monty'), who went on to become probably the best-known British general of the Second World War, had a complex personality. He was outwardly very assured, often abrasive towards those around him and keen on self-promotion – he always made himself available for photographers and film cameramen, no matter how busy and tense the situation was. But he was also something of a loner who did not fit in to the relaxed cycle of officer life in the British Army and seldom socialised in the conventional way. He had been a staff officer in the First World War and had learned the vital importance of thorough, detailed planning before combat. In the inter-war period, he was a training instructor for many

years and by 1939 he was a highly regarded, all-round professional soldier. He fought well in France in May–June 1940 and
impressed Churchill with his aggressive spirit when he was
then put in command of a division on the south coast waiting
for the anticipated German invasion. But Churchill also found
him an awkward man to work with. In many ways, Monty was
the opposite of the sort of general Churchill liked. He was
cautious, calculating and believed that no offensive action
should be taken until an army had built up an overwhelming
superiority of men, guns and equipment and could be sure of
supremacy in the air. He was lucky to arrive in Egypt just as
that was happening. With the new Sherman tanks and field
artillery arriving from America, as well as substantial reinforcements, Monty soon enjoyed a superiority of roughly two
to one over the Germans and Italians of Rommel's Afrika
Korps.

But Monty was more than just an excellent planner. He elec-
trified his headquarters and had an almost instant impact on
the morale of the men. He believed in constant training and the
need for physical toughness. He soon had the troops taking
daily exercises and endlessly preparing for battle. He also
believed a general should live among his men, and his constant
visits and pep-talks soon engendered a new fighting spirit. It
was Monty who would lead the next decisive phase of the
desert war. If he failed, and Rommel broke through to Cairo
and the Suez Canal, Hitler's plan was for the Afrika Korps to
link up with the southern flank of the army in Russia and to
capture the vast oilfields of the Caucasus and Persia. The stakes
could not have been higher.

Brooke felt 'that at last he [Churchill] is beginning to take my
advice', as the decision to appoint Alexander and Montgomery
to their key positions at this crucial moment had largely been
his.[36] As a result, Churchill at last had the right team in place.

He could work well with Brooke as his leading military adviser. And Alexander was the right strategic Commander-in-Chief for the Middle East, and did not interfere with Monty, who was left to make his own operational plans for battle.

The Ultra intelligence coming through about the state of the Afrika Korps was now really making a difference. Montgomery was forewarned about Rommel's next attack at El Alamein along the Alam Halfa Ridge at the end of August, so he reinforced his troops there. Rommel was unable to break through. By September, more Ultra decrypts helped the RAF to locate and sink about one-third of all the cargo that was sailing across the Mediterranean to reinforce the Afrika Korps, and almost half the Germans' fuel supplies. Increasingly, the intercepted messages reported how hard pressed the Afrika Korps were. Both sides continued to face each other in the desert, but the 8th Army was rapidly growing in men, equipment and, under Monty's inspired command, in self-confidence.

On the moonlit evening of 23 October 1942, after a huge artillery barrage, Monty launched his great offensive at El Alamein. His 8th Army consisted of British, Australian, New Zealand, Indian and South African troops. Monty feigned an attack in the south of his forty-mile front, which Rommel believed was the real thing. But his main thrust came in the north. For several days there was heavy fighting with neither side gaining a decisive advantage. But slowly the weight of armour as well as the persistence and determination of Monty and his men began to count. After twelve days, Montgomery wrote in his diary, 'The dam has burst,' and Rommel ordered his Afrika Korps to retreat. Harried constantly by the RAF, Rommel orchestrated a brilliant withdrawal along the coastal road, escaping to fight another day. Churchill was elated by the victory. The Battle of El Alamein proved to be a turning point in the war. In total, 8000 Germans and 22,000 Italians were taken prisoner, while 35,000 Afrika Korps and 13,000 Allied soldiers were killed or wounded. Montgomery had shown that the German Army was not invincible and that an Allied victory was a distant but real possibility.

In a speech in London, Churchill at last had some good news to announce. 'I have never promised anything but blood, tears, toil and sweat,' he said. 'Now however, we have a new experience. We have victory – a remarkable and definite victory . . . This is not the end. It is not even the beginning of the end. But it is, perhaps, the end of the beginning.'[37] On 8 November, Anglo-American troops landed in force along the French North African coast in Operation Torch. This was the first truly combined operation of the war and the result of the joint planning that had begun in Washington eleven months before. It was soon evident that the landings had been a success. Rommel was now in full retreat westwards from Egypt and troops from the Torch landings would begin to move east towards Tunisia. On 13 November, Monty's troops entered Tobruk, the loss of which had so upset Churchill five months earlier. On Sunday 15 November, Churchill ordered the church bells to be rung across Britain. Two years earlier, this would have been the signal that the German invasion had started. Now it was a sign of victory. At long last Churchill's generals had delivered him a victory that mattered.

7

The Admirals

With the whole French seaboard from Calais to Bordeaux under Nazi control after the fall of France, the Germans lost no time in building U-boat pens along the Atlantic coast. The U-boat war now turned even deadlier. During the last six months of 1940, U-boats sank 471 ships in the Atlantic, a total of more than two million tons of shipping. In the early months of 1941, the losses continued to rise alarmingly. Churchill was very aware of the need to keep the sea-lanes across the Atlantic open, not only for the supplies of guns, tanks and ammunition that were now coming out of America's factories under Lend-Lease, not only for the reinforcements in troops and *matériel* from Canada and the rest of the Empire, but for the foodstuffs, metal ores, oil, chemicals and all the other imports that were essential for Britain's survival. This was a struggle that Britain, as a trading nation, could simply not afford to lose.

Looking back on the titanic struggle for the Atlantic, Churchill wrote: 'The only thing that ever really frightened me during the war was the U-boat peril. I was even more anxious about this battle than I had been about the glorious air fight called the Battle of Britain.'[1] This was a conflict that could be followed only on charts, and with the accumulation of statistics on a weekly or monthly basis. The U-boats attacked in so-called 'wolf packs', assembling at night once a convoy had been spotted and then attacking in numbers, overwhelming the convoys' defences and sinking merchant ships almost at will. Added to the U-boat menace were the activities of surface raiders like the *Scharnhorst* and the *Gneisenau*, German battle cruisers that between them sank or captured twenty-two ships in early 1941. And the long-range Condor aircraft could also spot convoys and bomb them from the air. 'How willingly would I have exchanged a full-scale attempt at invasion for this shapeless, measureless peril, expressed in charts, curves and statistics!' wrote Churchill.[2] In February 1941, the command centre for the 'Western Approaches', as the Atlantic sea-lanes into Britain were called, was moved to Liverpool and Admiral Sir Percy Noble was put in command. The following month, after some particularly bad losses, Churchill met with Admiral Pound, the First Sea Lord, and told him that this struggle had to be given priority over everything else. Churchill gave a new name to this campaign: the 'Battle of the Atlantic'. As before, his use of words created a new battle-cry. By delineating the struggle for the Atlantic as a battle, just like the Battle of France or the Battle of Britain, Churchill concentrated minds, rallied government departments and focused the public's eye on a key matter of survival.[3]

A new committee dedicated to the Battle of the Atlantic was set up. For several months it met weekly, then fortnightly, to review all the information and to come up with new ideas and policies. It was chaired by Churchill himself and consisted of the rest of the War Cabinet, the naval and air Chiefs of Staff, other key ministers, and leading scientists such as Lord

Cherwell and Professor Blackett, who brought the skills of
Operational Research to bear on the subject. Churchill wrote a
directive listing thirteen points to get the new committee going.
It was a magisterial document calling for an offensive against
the U-boat menace – as ever, Churchill wanted to take the attack
to the enemy. It helped to galvanise the minds of all the key
players in this new battle. It was Churchill at his aggressive,
motivational and coordinating best.[4]

At the beginning of the war, the only real weapon Britain had
against the U-boat menace was Asdic, which the Admiralty had
developed in the inter-war years. This was a system that sent
out sound pulses that bounced back when they hit something,
a bit like an underwater version of radar. At the Admiralty,
Churchill had been keen on Asdic. But its range was limited
and it was unreliable. It could be disrupted by water turbulence
caused by the wakes of ships or by depth charges. However, by
1941, new measures against the U-boats had begun to appear.
One of these was a radio direction finding system that picked
up the enciphered messages of U-boats as they sent signals to
their headquarters near L'Orient in France. These messages
could not be deciphered, but that didn't matter as the objective
was simply to locate where the signals were coming from. If
two or more radio stations picked up the signals then they
could 'fix' on the position of the U-boat. This could be done at
long range and with impressive accuracy. The system, called
'High Frequency Direction Finding', or HF/DF, was nicknamed
'Huff Duff'. The wizards were beginning to influence the war
at sea as well.

Other improvements in 1941 included readjusting the depth
at which depth charges ignited, the use of new forms of radar
and the deployment of fast-moving destroyers as convoy
escorts. Hitler also inadvertently helped the Allies in the Battle
of the Atlantic by ordering Admiral Karl Doenitz, the U-boat

Commander-in-Chief, to redeploy some of his vessels to the Mediterranean. But a decisive breakthrough came in May and June 1941, when the Royal Navy captured the German *U-110* submarine intact, along with two weather ships, and seized all of their Enigma code books. With them the code-breakers at Bletchley Park were at last able to break the German Navy's encrypted messages. By deciphering key signals to and from the 'wolf packs', it was possible to confirm their locations and their intentions. The convoys could then be routed away from the waiting U-boats. This dramatically reduced shipping losses, from an average of about a quarter of a million tons per month in the first six months of 1941 to roughly half of that, and by the end of the year even less. This was a victory for this phase of the Battle of the Atlantic. But it would not last long.

Another intense battle in the Atlantic attracted far more attention. After an intelligence tip-off from local agents, an RAF photographic reconnaissance Spitfire set out from Wick in the north of Scotland on 21 May 1941 and headed for the Norwegian fjords around Bergen. The pilots of these reconnaissance aircraft were true heroes. They flew without guns, sometimes for hours at a time, over enemy-occupied territory. Speed was their only weapon, and if fighters came after them it was only their ability to get away quickly at high altitude that saved them. On this afternoon, the pilot spotted a group of German warships, went down to have a look, then saw another two warships, one very large, about to sail. When the photographs were processed and examined, it transpired that the pilot had found the *Bismarck* and its support cruiser the *Prinz Eugen* about to head into the Atlantic. The *Bismarck* was the pride of Hitler's navy and its newest addition. It was the biggest German battleship ever built and boasted eight fifteen-inch guns. At 45,000 tons it was also heavier than the largest British battleship, but just as fast. If it got among the

Atlantic convoys it could cause mayhem. One of the most famous pursuits in naval history now began. The Admiralty ordered ships from Scapa Flow and from across the Atlantic to converge on the ice-bound stretch of water between Greenland and Iceland known as the Denmark Strait, where it was reckoned the *Bismarck* would attempt to pass into the ocean.

Churchill was at Chequers that weekend following the campaign that was raging on the island of Crete. At seven o'clock on the morning of Saturday 24 May, he was woken with dreadful news. In the first engagement with the *Bismarck*, HMS *Hood* had been sunk. *Hood* was an ageing battle cruiser but still the pride of the Royal Navy, one of its latest battleships. An inquiry found that a single shell from the *Bismarck* had penetrated the *Hood*'s ammunition store and the ship had exploded. Only three men from its crew of fifteen hundred survived. Another British battleship, HMS *Prince of Wales*, was also substantially damaged by the giant shells of the *Bismarck* and its bridge was put out of action. However, although it didn't realise it, the *Prince of Wales* had managed to hit the *Bismarck*, which had slowed it down.

The signal went out 'Sink the *Bismarck*' and all available warships in the Atlantic were ordered to join the pursuit. Battleships, cruisers, aircraft carriers and destroyers now raced to converge in the seas to the south of the Denmark Strait. On the evening of the 24th, Fleet Air Arm Swordfish aircraft flying from HMS *Victorious* spotted the *Bismarck*. The Swordfish was a biplane with canvas wings held together with bracing wire and was known by its crews as the 'Stringbag'. Although it looked like a leftover from another era, it was sturdy and reliable and carried a torpedo. One of these scored a direct hit under the bridge of the *Bismarck*. Then, at this critical moment, the pursuing cruisers lost radar contact with their prey. For a whole day, the *Bismarck* could not be found, despite a frantic search. In what direction was she now heading? On the 26th, a

Catalina flying boat from Lough Erne in Northern Ireland made a lucky identification. The *Bismarck* was heading for the port of Brest for repairs. It was about 700 miles out to sea. Soon, Swordfish from HMS *Ark Royal* found the battleship and more of their torpedoes hit home. Admiral Tovey, Commander-in-Chief of the Home Fleet, was aboard HMS *King George V*, which was now in hot pursuit along with HMS *Rodney*. But these two battleships were running low on fuel and it looked like they might have to give up the chase. Churchill ordered Admiral Pound to send a cable saying, '*Bismarck* must be sunk at all costs', even if one of the British battleships had to tow the other back afterwards.[5]

By now, the damage to the *Bismarck* had caused her rudder to jam and she could sail only hopelessly round and round in a circle. On the morning of the 27th, *King George V* and *Rodney* finally closed in for the kill. The *Bismarck*'s heavy guns were still firing and caused damage to the *Rodney*, but the German ship was a sitting duck and a combination of shells and torpedoes ultimately did for her. Admiral Lutjens, the German commander, sent a final signal: 'Ship unmanoeuvrable. We shall fight to the last shell. Long live the Fuehrer!' When the *Bismarck* went down, all but about 120 of its 2000-man crew, including Lutjens, went down with it. It was a victory for the Royal Navy, although with the loss of the *Hood*, a costly one. But it did deter the *Tirpitz*, another German heavy battleship, from ever putting to sea. Instead, it spent much of the war sheltering in the fjords of Norway.

Churchill had great experience of civilian command of the Royal Navy, going back thirty years to when he was appointed First Lord of the Admiralty in 1911. Back then, he had found the men who commanded the senior service very conservative and suspicious of change. Now, in the Second World War, armed with the confidence and conviction that he knew best, he was determined to have his say in operational matters. This led to considerable strain in his dealings with the admirals, just as it did with the generals when Churchill interfered in army affairs. But Admiral Pound knew Churchill well, having worked with him since September 1939 as First Lord of the Admiralty and then as Prime Minister on the Chiefs of Staff committee. He knew how to put up with him and they formed a good partnership. In December 1940, he told Admiral Cunningham, 'The PM is very difficult these days, not that he has not always been. One has however to take a broad view as one is dealing with a man who has proved to be a magnificent leader, and one just has to put up with his childishness as long as it isn't dangerous.' Later, Pound said to A.V. Alexander, Churchill's replacement as First Lord, 'At times you could kiss his [Churchill's] feet. At others you feel you could kill him.'[6]

Admiral Sir Andrew Browne Cunningham, universally known as 'ABC', commanded the Mediterranean Fleet until late 1943. He was the most successful British naval commander of the Second World War and has been called the greatest admiral since Nelson. He was behind the air attack on the Italian fleet at Taranto and masterminded the Battle of Matapan in March 1941, when he pursued his attack upon the Italian fleet at night and three Italian heavy cruisers were sunk for no British losses. He had a natural instinct to close with and destroy the enemy at every opportunity. Even when organising the evacuation of Greece and then Crete in April and May 1941, without air cover, he told his sailors that the 'Navy must not let the Army down'. But even he was subject to Churchill's detailed directives. Cunningham found these undermining, often believing that they cast doubt on his intention to take the fight to the enemy. Mostly he bit his lip and carried out his duties as best as he could. But after an argument over the blockading of Tripoli in April 1941, he admitted he was 'beginning to feel seriously annoyed' at being told how to do his job. He felt Churchill's instructions could have 'lost the whole fleet' if circumstances had turned out differently.[7]

But Churchill never maintained his criticism of anyone for long. Not only did he warmly congratulate Cunningham but he also offered him several promotions. Churchill was good at prioritising and at seeing the overview. He was brilliant at making things happen and refusing to take 'no' for an answer. And, of course, he had to balance all the competing demands for limited war supplies and to set overall priorities which were bound to annoy whichever party lost out. But he could go too far, and his micro-instructions turned potentially friendly military chiefs against him. At times, his interference even risked the lives of soldiers, sailors and airmen. His commanders were right to resist some of his demands and to argue back. A lid was kept on these arguments during the war, but some of the frustration came out in memoirs published after the war.[8]

We have seen how Hitler's invasion of the Soviet Union, the Japanese attack on Pearl Harbor and Germany's declaration of war on the United States truly created a world at war. The pressure on the Royal Navy, fighting in so many oceans, was now immense. Three days after Pearl Harbor, on the morning of 10 December, Churchill was still reading his boxes in bed when the phone rang. It was Admiral Pound. Churchill later wrote, 'His voice sounded odd. He gave a sort of cough and gulp and at first I could not hear him quite clearly.' The news he brought was shocking. HMS *Prince of Wales* and HMS *Repulse*, two battleships that had been sent to the Far East before the attack on Pearl Harbor in a bid to deter the Japanese, had been sunk by Japanese aircraft. Churchill was distraught. He had sailed on the *Prince of Wales* only a few months before to meet Roosevelt in Newfoundland. He knew many members of the crew well. 'In all the war I never received a more direct shock,' he wrote later. 'As I turned over and twisted in bed the full horror of the news sank in upon me.' Coming so soon after the disaster of Pearl Harbor, the sinking of the two ships meant there were now no Allied capital ships left in the Pacific. Churchill reflected: 'Over all this vast expanse of waters Japan was supreme, and we everywhere were weak and naked.'⁹

The sinking of the *Prince of Wales* and the *Repulse,* along with the pursuit of the *Bismarck,* proved how vital air power was at sea. The *Bismarck* had been found and critically injured by naval aircraft. The *Prince of Wales* and the *Repulse* had left their air cover behind and had been sunk by Japanese bombers. Then, the Battle of Midway in June 1942 was fought largely between aircraft flying from carriers. It would be one of the turning-point engagements of the Pacific naval war, with US naval aviators inflicting the loss of four aircraft carriers on the Japanese Imperial Navy. But still some in the Admiralty refused to recognise or accept the change that was taking place in naval warfare. The day of the heavy battleship operating as a floating steel platform for its giant guns was passing. The future, particularly in the Pacific, lay with aircraft carriers. And the United States soon began a huge aircraft-carrier building plan. By the end of the war in the Pacific, US shipyards had built twenty-four vast, 27,000-ton Essex-class carriers. They would slowly restore US naval supremacy across the 'vast expanse' of the Pacific Ocean.

But it would be some time before the disastrous defeats in the Far East could be avenged. More than any other incident, the humiliating loss of Singapore showed up British military weakness in Asia. Singapore is a small island at the tip of the five-hundred-mile-long Malayan peninsula. In the 1920s, it had been selected as the site for the construction of a huge naval base and a fortress with heavy guns as the centrepiece of British power in the region. Giant defences with heavy artillery pointing out to sea had been constructed to defend the city and the trading hub from seaborne attack. But Singapore's land defences were left almost non-existent, and it was from the land, down the Malayan peninsula, that the Japanese came. Without doubt, this was a disastrous lack of foresight and planning, a blunder of historic proportions. Arguments have raged for years over who was to blame. Certainly Churchill was partly responsible, as he was Chancellor of the Exchequer in the late 1920s and kept a tight rein on spending when the Singapore fortress was being built. And from 1939 he was first a senior member and then the leader of a government that failed to build up proper defences. But a war fought simultaneously with Germany, Italy and Japan had simply never been envisaged, and all planning since September 1939 had focused on Europe and North Africa. Despite his own culpability, though, Churchill was furious when he was told about the lack of defence against a land attack. He wrote that he was 'staggered', and in January 1942 he fumed to the Chiefs of Staff: 'What is the point of having an island for a fortress if it is not to be made into a citadel?' He continued that the provision of adequate land defence 'was an elementary peace-time provision which it is incredible did not exist in a fortress which has been twenty years building . . . How was it that not one of you pointed this out to me at any times when these matters have been under discussion?'[10]

Thousands of Allied land troops packed into Singapore after a demoralising and exhausting retreat down the length of Malaya. General Arthur Percival was in command. Churchill instructed him to stand and fight, and said that the unit in the forefront of the battle, 18th Division, 'has a chance to make its name in history . . . The honour of the British Empire and of the British Army is at stake.'[11] But the Japanese onslaught rolled relentlessly forward, and the dispirited defenders found they were in a hopeless situation. Water supplies were running out and the one million civilian inhabitants of the city were desperate. Ammunition and fuel reserves were also disastrously low. Churchill discussed the situation with Brooke and agreed that it was unrealistic to insist on holding out any longer. Through the regional commander, they gave Percival permission to cease resistance when he saw fit. On Sunday 15 February, Percival and his commanders marched out with white flags to signal the unconditional surrender of the fortress. Sixty-two thousand men went into captivity and began years of horror and torture in Japanese prisoner-of-war and work camps where half of them would die. Churchill wrote later that it was 'the worst disaster and largest capitulation in British history'.[12]

That evening Churchill announced the fall of Singapore on the BBC. It was an emotional speech in which Churchill sought to review the progress of the war by asking: 'Are we up or down?' He insisted that with both the United States and the Soviet Union as allies, victory was ultimately certain. But he admitted that the position for Britain was severe, and that the country faced hard struggles on every front. He ended with a typical rhetorical flourish by saying that under the shadow of a disastrous defeat, it was now one of those moments

> when the British race and nation can show their quality and their genius. This is one of those moments when it can draw from the heart of misfortune the vital impulses of victory . . . We must remember that we are no longer alone. We are in the midst of a great company . . . So far we have not failed. We shall not fail now. Let us move forward steadfastly together into the storm and through the storm.

Interestingly, the following day, Harold Nicolson wrote in his diary that Churchill's broadcast was not liked: 'The country is too nervous and irritable to be fobbed off with fine phrases.'[13] At this point of the war, it seems that the Churchillian magic was no longer working.

The global war was exacting its price. Neither the Royal Navy nor the British Army was capable of fighting effectively on so many fronts over such vast distances. Furthermore, Churchill never really empathised with the war in the Far East. He had never studied Japanese society and had very little understanding of Japan's military culture. He had a tendency, which he shared with his military chiefs, constantly to underestimate this new enemy's strength. Moreover, he had never visited Australia, and so never fully grasped the Australian fear that they were now at threat from the Japanese advance as it rolled mercilessly on across the region. He just couldn't see why the Australians felt the need to withdraw their troops from the Middle East.

For Churchill, the key front in the Far East was the defence of India, the jewel in the imperial crown, a place of course very close to his heart after his service there as a young cavalry officer and his championing of imperial interests in the country ever since. Defending India meant defending Burma, and so for Churchill this was the critical aspect of the war against Japan at this stage. In January 1942, he argued that Singapore should be abandoned in order to build up the defence of Rangoon, the Burmese capital. But to his regret, under pressure of other events he did not push for this. Burma fell to the Japanese in the spring. Sensing this moment of British weakness in Asia, a few months later the Congress Party in India launched a series of demonstrations and riots against British rule. With Germany advancing into southern Russia, potentially threatening the northwest frontier of India, and with Japan strident to the east, it looked to many that British rule in India might be entering its final days.

Back in the Atlantic, a fatal blow was struck when the German Navy changed the configuration of its Enigma machines. When the *Bismarck* had sailed, five supply ships had been dispatched to various points in the Atlantic so the mighty battleship could refuel at sea. Bletchley Park deciphered their codes, learned their locations, and passed on the information. Once the *Bismarck* had been sunk, four of the supply ships were tracked down and also sunk. The fifth was spotted by chance and sunk. Admiral Doenitz, the commander of the U-boat fleet, began to get suspicious that the British were decoding his Enigma signals. Then *U-570* was captured by a British destroyer south of Iceland, and it was feared that the Royal Navy might have seized the U-boat's Enigma machine and code books (in fact, this time, it had not). An investigation into possible code-breaking was carried out by Captain Stummel, an experienced naval signals officer with an almost comic-book Prussian appearance straight out of central casting with a glass eye and a limp. Stummel concluded that it was impossible to break Enigma, but Doenitz still wanted to tighten up the whole system. As a result, a fourth rotor blade was added to the Enigma machines, meaning that the code permutations were increased by a factor of twenty-six. The new system came into effect on 1 February 1942.[14]

Overnight, this resulted in a complete blackout at Bletchley Park of the German naval codes. The code-breakers called the new, seemingly impenetrable code 'Shark'. It created a major crisis in the Battle of the Atlantic. And it coincided with a glorious period for U-boats along the the eastern seaboard of the United States. Ignoring the lessons that had been so painfully learned by the British about the need for convoys, single ships continued to sail up and down the US coast and there was not even a blackout in the port towns. The U-boat crews enjoyed the easiest hunting of the war in what they called 'the happy time'. They sunk 31 American ships in January 1942, 69 in

February, and the numbers continued to rise over the next six months until sensible precautions and a system of integrated convoys were brought in.

From the summer of 1942, the number of ships lost crossing the Atlantic also grew to frightening levels. In the six months following the launch of the new Enigma system, U-boats sank about five hundred ships in Atlantic convoys, getting on for three million tons of shipping (in addition to all the ships lost along the eastern seaboard of the United States). At this rate many more ships were being sunk than could be built to replace them. And quite apart from the terrible loss of life, so many supplies were ending up at the bottom of the Atlantic that Britain risked running out of rations and the raw materials necessary to fight the war. By now, the Germans had also broken Britain's naval cipher and their code-breakers were picking up the routes of and instructions to the convoys as they set out from North America. The Admiralty stepped up the pressure on the code-breakers at Bletchley to crack the new Enigma. In Hut 8, where Alan Turing was based, they worked day and night. But they were operating almost blind and made little progress.

By November 1942, Churchill was once again facing a potentially catastrophic situation in the Atlantic. A second committee was formed, the Cabinet Anti-U-boat Warfare Committee. Like its successful 1941 predecessor, it was chaired by Churchill and consisted of senior ministers, the service chiefs and top scientists, including Robert Watson-Watt, Lord Cherwell and Professor Blackett. The presence of the scientists was key, and the committee increasingly brought scientific thinking to bear on the Battle of the Atlantic. A series of important developments soon followed. U-boats travelling on the surface had developed the technology to pick up traditional radar signals and therefore knew when they had been spotted by Allied aircraft. They could then dive below the surface and evade an air attack.

However, new, more sophisticated forms of short-wave radar (made possible by the development of the cavity magnetron at Birmingham) could not be detected by the U-boats. This dramatically increased the ability of patrolling aircraft to attack and sink U-boats, especially in the Bay of Biscay as they set out for or returned from their missions. A new searchlight known as the 'Leigh Lamp' was fitted to the aircraft so that they once they had tracked a U-boat using the new short-wave radar they could then attack at night. A new class of frigate was developed and dispatched in groups of four as independent flotillas to hunt down U-boats. Several escort carriers were built so that convoys could be accompanied by aircraft, and this extended the range at which U-boats could be hunted down. The boffins in Operational Research did their sums and noted that the U-boats were obviously limited by the number of torpedoes they could carry. So the loss rate was proportionately lower in a big convoy than in a small one. Accordingly, the convoys were enlarged. All of this helped. But some masterstroke was still needed to overcome Shark.

In fact, the lucky breakthrough had already occurred on 30 October, but it took some time for its impact to be realised. Early that morning, a Sunderland flying boat patrolling over the eastern Mediterranean picked up a radar contact with what it reported was possibly a submarine. A group of the latest class of Royal Navy destroyers was on patrol in the area and rushed to the coordinates supplied by the plane. The ships' Asdic underwater sonars picked up the presence of a U-boat and they began to launch depth charges. Throughout the afternoon and evening, the destroyers circled and dropped more explosives into the sea. The U-boat dropped to its maximum depth, below five hundred feet, but the charges continued to come. The stench of sweat, fear and diesel fumes inside the submarine grew worse and worse. At about 10 p.m she was struck despite her depth. The captain decided he had no alternative but to surface. When *U-559* came up, one of the circling destroyers, HMS *Petard*, opened fire and holed the conning tower. The U-boat crew abandoned ship and leapt into the sea. The *Petard*'s captain realised he might be able to capture the U-boat and tow it to port. So a boarding party was quickly sent across and clambered on to the U-boat. In this party was Lieutenant Anthony Fasson, who knew the importance of secret radio documents held on each U-boat. He descended into the U-boat, where the lights were still on. It was holed badly and taking in water. In the captain's cabin, Fasson broke open a cabinet, took a key and opened a drawer where he found books and papers. With the help of another member of the boarding party, Colin Grazier, Fasson brought up several piles of these documents to the conning tower. On their third trip back down to the captain's cabin, the U-boat suddenly started to sink. The officers on the conning tower managed to get off with most of the documents, but Fasson and Grazier were caught inside by the inrush of water and went down with the U-boat. They were both awarded posthumous George Crosses for their heroism.[15]

When the code books seized by Fasson and Grazier from *U-559* finally arrived at Bletchley Park, they provided the missing link. In the language of cryptology, the code-breakers were now able to find their cribs and kisses, and use their bombes. On the morning of 13 December, a cry went up from Hut 8: 'It's out!' They had cracked the four-rotor Enigma cipher. They could now read Shark. Within days, they had identified the positions of fifteen U-boats. Within weeks, the process of diverting convoys away from the waiting wolf packs could be resumed. By January and February 1943, shipping losses had dropped to half of what they had been before the breakthrough. It was the decisive moment in the Battle of the Atlantic. But the situation was to get a lot worse before it finally got better.

In the final months of 1942, the alliance of interests that had
united Britain and the United States began to diverge. Admiral
King, the head of the US Navy, and General Douglas MacArthur,
the charismatic commander who had led the US garrison in the
Philippines, both argued forcefully that more resources were
needed for the war in the Pacific. This challenged the Germany
First principle that Roosevelt had agreed with Churchill soon
after Pearl Harbor. And General Marshall, the US Army Chief
of Staff, who had gone along with the idea of Operation Torch
and the landings in North Africa, now argued that an invasion
of northern France in 1943 should be a higher priority than continuing the war in the Mediterranean. Roosevelt suggested to
Churchill that they should meet with Stalin to review strategy
for the year ahead. But Stalin announced that at this critical
point in the Battle of Stalingrad, he could not leave Moscow.
Churchill and Roosevelt decided to go ahead without him, and
plans were drawn up for the two men to meet at Casablanca in
Morocco. The summit was fittingly given the code name
'Symbol'.

On 12 January 1943, Churchill set off in a Liberator bomber
for the nine-hour flight to Morocco. He travelled with his usual
small entourage, his doctor, Sir Charles Wilson, John Martin
from his private staff, his detective Tommy Thompson and his
valet Frank Sawyers. It was, as ever, an uncomfortable flight,
and at one point Churchill awoke to find a pipe overheating
and feared that it would ignite petrol fumes in the cabin. But
they arrived safely and the warm winter sun of Morocco
brought welcome relief. 'Bright sunshine, oranges, eggs and
razor-blades,' wrote Martin, all four items being in short supply
in Britain at the time.[16] A small, discreet hotel had been taken
over on the coast to the north of Casablanca. Churchill and
Roosevelt had their own private villas, and the whole com-
pound was sealed off and guarded by US Marines. Attending

the conference were the British and American Chiefs of Staff, who met separately each day and then usually had two daily meetings together as the Combined Chiefs. The British had prepared their position and their arguments well. Churchill and his Chiefs of Staff had talked through all the key points and were now in agreement. The Americans were less well prepared and indeed had not even agreed on a unified position.

The conference formally opened on 13 January and the divisions between the military leaders immediately became evident. The Americans wanted to plan for an invasion of Northern Europe. The British wanted to concentrate on victory in the Mediterranean and considered an invasion of France in 1943 too great a risk. They argued for postponement until 1944, when more resources would be available and the U-boat threat should have been defeated. For several days, the two sides were locked in disagreement as to their strategy for the following year. General Brooke recorded in his diary the 'very heated' meetings with the Americans which seemed to be making 'no progress', and on the fifth day he wrote: 'A desperate day! We are further from obtaining agreement than we ever were!'[17] The British chiefs feared the Americans were already reallocating resources from Europe to the war in the Pacific. The Americans thought the British were obsessed with the Mediterranean for reasons of self-interest and traditional imperial concerns to retain the Suez Canal as a link to India.

While all of this was being thrashed out, Churchill and Roosevelt were getting on famously. The President seemed more relaxed than his Chief of Staff Marshall on the issue of the Mediterranean strategy. In return, Churchill responded to American fears that Britain would pull out of the war as soon as Germany was defeated by pledging to continue the fight against Japan. The warm friendship between the Prime Minister and the President blossomed in the winter sunshine.

And Churchill was in his element, with his military chiefs around him, seeing Roosevelt regularly, and being at the heart of a global war machine. Harold Macmillan, who joined the conference a few days after it had started, wrote: 'I have never seen him in better form. He ate and drank enormously all the time, settled huge problems, played bagatelle and bezique by the hour, and generally enjoyed himself.'[18]

On 18 January, the Combined Chiefs finally reached an agreement on strategy. Many of the Americans felt comfortable dealing with Field Marshal Sir John Dill who had been head of the British Mission in Washington for over a year and who was on excellent terms with Marshall. He played a central part in bringing both sides together. The Combined Chiefs presented their findings to Churchill and Roosevelt, who enthusiastically endorsed the new plans. Most of what the well-prepared British team had set out to achieve had been accepted. Defeat of Germany was still the first priority. Overcoming the U-boat threat and winning the Battle of the Atlantic was the leading objective. The Mediterranean strategy was endorsed, with agreement to invade Sicily once the Allies had finally expelled the Afrika Korps from North Africa. America agreed to support a British invasion of Burma later in the year. In Europe, Operation Bolero, the build-up of US troops in Britain, was to go ahead with maximum urgency. And the bombing offensive against Germany was to continue with new vigour. Later in the conference, Roosevelt added the condition that the Allies would accept only the 'unconditional surrender' of both Germany and Japan. This was agreed by Churchill on the spot and became the ultimate Allied war objective.

For the British planners, the summit was an enormous triumph. A joint strategy had been agreed by negotiation between the military chiefs, and the British had persuaded the Americans of the strengths of their case. It had not been handed down

by edict from the US Commander-in-Chief and the British
Prime Minister. In many ways, Casablanca represented the high
water mark of British influence over the planning of the war.
During 1943, the vast scale of the US war effort would over-
take Britain and leave Churchill and the British chiefs as very
much the junior partner in the alliance. But now, fittingly,
Churchill and Roosevelt left Casablanca and travelled together
to Marrakech, a city that Churchill had fallen in love with years
before on holiday. The Prime Minister climbed to the roof of the
villa that had been requisitioned for them to watch the beautiful sunset over the snow-capped Atlas mountains. He insisted
that Roosevelt must see it too, and the disabled President was
literally carried to the roof by two of his staff. Together they
watched as the light magnificently changed colour in front
of them. Churchill murmured that 'this was the most lovely
spot in the whole world'. They concluded their business over
dinner by drafting various communiqués, including a summary
for Stalin of what had been agreed. And then they sang songs
together.[19] The following morning, the President left for the
United States and Churchill took a rare break. He painted the
view of the mountains from the roof that he and Roosevelt had
so admired the evening before. It was the only painting Churchill
produced during the whole war.

Throughout 1942, U-boats had been launched at an average
of eighteen vessels each month. By early 1943, Doenitz had
about a hundred operating in the North Atlantic. There was a
zone in the centre of the ocean that aircraft could not reach from
either the United States or Britain. Inside this gap the U-boats
continued their killing spree. Churchill came under considerable pressure from the Admiralty to divert the big, four-engined
RAF aircraft like the Liberator from Bomber Command to
Coastal Command in order to fly sorties over the middle of
the Atlantic to protect the convoys. But he and his advisers

remained convinced that in the absence of a second front to satisfy Stalin, the bombing offensive was the principal method available to them to destroy the Nazi war machine. The Admiralty's request for the long-range aircraft was turned down. Instead, more bombing raids were made on the U-boat pens along the French coast. But the concrete roofs were so massive that despite literally thousands of bombing missions, not a single bomb ever penetrated to the U-boat docks below.

In March 1943, the sheer number of U-boats meant that shipping losses increased to over 600,000 tons, an appallingly high figure. Britain could not survive for long with this rate of loss. The Admiralty calculated that in the first twenty days of this month, communications between the Old World and the New nearly broke down completely. Stephen Roskill, the official naval historian, wrote: 'in the early spring of 1943 we had a very narrow escape from defeat in the Atlantic . . . had we suffered such a defeat, history would have judged that the main cause would have been the lack of two more squadrons of very long range aircraft for convoy escort duties'.[20] Churchill's adherence to the bombing offensive very nearly led to defeat in the Battle of the Atlantic.

The advances in science came to the rescue just in time. All of the developments that had been taking place at last began to kick in. With Bletchley Park now able to read the German naval signals again, with the new short-wave radar systems picking up more and more U-boats on the surface, and with the new escort groups in place, the hunters became the hunted. Aircraft patrolling the Bay of Biscay were spotting and sinking U-boats as they set off on or returned from their missions. In March and April, twenty-seven U-boats were sunk in the Atlantic, more than half by attack from the air. In May, forty were sunk, eighteen of them by air attack. In total, this amounted to over 50 per cent of the available U-boat fleet. No force could sustain this rate of loss. On 22 May, Doenitz called off the campaign and ordered the withdrawal of his U-boats from the North Atlantic. The convoy code numbered SC130, which reached Liverpool on 23 May, was the last to be seriously menaced by U-boats.

The sea-lanes from North America to Britain had been cleared of the U-boat menace. It was as vital a turning point as Stalingrad or El Alamein. It was inconceivable that an invasion of Europe could have been mounted without mastery of the North Atlantic. And the great shipyards of North America were now working at full throttle, many producing what were known as 'Liberty Ships'. These were basic transport vessels that could be prefabricated in several parts. Henry Kaiser, the shipbuilding magnate, realised that speed of construction was of the essence, and that as the shipyards expanded much of the available labour was going to be unskilled. The genius of Kaiser's idea of prefabricating the ships was that many of the parts could be produced at inland factories, then transported by railway to shipyards on the coast for assembly. A giant ship's hull could be designed in a series of subsections, and new techniques like electric arc-welding were used to bond the plates together. This was faster than riveting and used less steel. More than 300,000 men and women were employed in the US shipyards up and down the east and west coasts. Many of them had never been near a shipyard in their lives before. Just as 'Rosie the Riveter' became the media darling of the aircraft manufacturing plants, so 'Wendy the Welder' became the generic name for women workers in the traditionally male-dominated world of heavy engineering and shipbuilding. The whole manufacturing process of the Liberty Ships was one of the most remarkable examples of mass production ever achieved. The speed at which these basic cargo ships could be assembled and launched got faster and faster. The first of the Liberty Ships took about 150 days from laying the keel to launch, itself impressively fast in shipbuilding terms. As techniques improved and systems became even more efficient, the time came down to fifty days. Then the American media got interested and a friendly rivalry was set up between the shipyards. Average production time fell to just ten days, and the fastest of all, the *Robert E. Parry*, was built at a yard in California in November 1942 in an incredible four days and fifteen hours. This speed of production was exceptional but in total 2700 Liberty Ships were manufactured during the war.

The U-boat threat was defeated by the application of scientific ideas at sea and by the revolutionary concept of mass-producing ships in America. As Professor Blackett put it: 'the anti-submarine campaign of 1943 was waged under closer scientific control than any other campaign in the history of the British Armed Forces'.[21] In July 1943, a key milestone was passed when the number of ships being built and launched exceeded the volume of shipping being lost. The War Lab had played its part and the Battle of the Atlantic had been won.

8

Bombing

On the night of 19 March 1940, a group of twenty Hampdens and thirty Whitleys of Bomber Command launched the first RAF bombing raid of the war on Germany. Their target was the German seaplane base of Hornum on the island of Sylt, a few miles west of the Danish border. From here, Luftwaffe seaplanes had been dropping mines in the North Sea. There was great excitement among the bomber crews who had spent the first few months of the war doing nothing more than dropping propaganda leaflets over Germany. Such was the fascination with this first bombing raid that late that night in the House of Commons, Prime Minister Neville Chamberlain interrupted the session to announce that RAF bombers were obliterating the German airbase. Members cheered. The newspapers responded with a brace of headlines: 'Hangars And Oil Tanks Ablaze' and 'Night Sky Lit Up'. At last the RAF was dishing it out to the Nazis.

However, when the first aerial photographs of Hornum came back on the day after the raid they told a very different and rather surprising story. The photo interpreters could find no damage whatsoever to the hangars. The Heinkel seaplanes were still on their slipways and the oil tanks were all extant. The interpreters examined the photos over and over again. They considered the possibilities. Had the Luftwaffe managed a remarkable overnight feat of clearing up the destruction? Slowly, they came to an inescapable conclusion. Despite the enthusiastic post-flight debriefings of the crew, who told of blazing hangars and workshops below, the RAF bombers had entirely missed their targets. As a propaganda coup, the Germans invited neutral journalists to visit the base and see for themselves that no damage had been done. But this went unreported in Britain.

The RAF had learned its first depressing lesson about bombing at night. With the rudimentary navigation techniques then available, it was almost impossible to fly for seven or eight hours across a blacked-out and hostile Europe and hit the correct target. Navigation was mostly by sight, identifying landmarks below and charting a course from them. Obviously, cloud cover made this difficult. And bombers flew at high altitude to be above anti-aircraft defences and searchlights; dropping down low to identify a landmark brought obvious risks. Occasionally, crews flying above the clouds could use a sextant to read the stars as a guide to their location, but this took time and required flying on a consistent and level straight line for longer than most pilots felt comfortable doing. From the start of a flight, when an aircraft charted its course, the wind could blow it astray and evasive action like weaving could take it further off course. This was why the Germans had developed their network of *Knickebein* beams to guide bombers to their targets. But the RAF had nothing like this, so the consequences of its early bombing missions were mostly embarrassing.[1]

As soon as Churchill became Prime Minister and the war entered its critical phase, there was much to be done despite the difficulties. RAF bombers were assigned targets of rail junctions and airbases, and set off in their slow-moving aircraft to try to hit them. Mostly they failed. One RAF bomber in late May 1940 was sent to bomb a German airfield in Holland. It hit an electrical storm over the North Sea and became hopelessly lost. Eventually, totally confused, the crew identified what they thought was the Rhine estuary, found what they believed to be the target airfield, dropped their bombs on it, and returned home. When they found themselves over Liverpool, they realised something had gone dreadfully wrong. They had mistaken the Thames estuary for the Rhine and had bombed an RAF airfield by mistake. Bomber crews usually carried out their missions using 'ETA', dropping their bombs at the Estimated Time of Arrival over their target, but this meant they were often not just miles off their targets, but were sometimes tens or even hundreds of miles off course.[2]

Churchill, as ever, closely followed the progress of the bombing attacks on Germany and raised a host of questions and suggestions. Within five days of becoming Prime Minister he authorised a raid on the Ruhr industrial district. He asked what were the types of bombs in use and what improvements could be made to the aircraft? It will be remembered that he had orchestrated the first bombing attacks by primitive canvas and wire aircraft in the First World War, and he had been the minister in charge of the Royal Air Force after the end of that war. During the inter-war years, the RAF needed to find a role to secure its survival as a force that was independent of the army and navy. It did this by emphasising its ability to strike at the enemy. There were two elements to this: to undermine the ability or will of the enemy nation to fight by hitting their homeland (this was known as 'strategic bombing'); and to hit the enemy's armies, ammunition dumps, supply lines and operating ability in the field (known as 'tactical bombing'). This was the era when it was believed 'the bomber will always get through', as Stanley Baldwin had said, so maintaining a strong bomber force was intended to deter an enemy from making a first strike. Air Marshal Sir Hugh Trenchard, Chief of the Air Staff, preached this doctrine and largely defined the shape of the inter-war RAF, which had twice as many bombers as fighters. In the final years of peace, it was realised at the eleventh hour that the RAF's fighter capability had to be rapidly built up to defend Britain from enemy bombers. It was ironic that the first great success of the RAF in the war was not in the offensive bombing of German military targets, but in the purely defensive action of Sir Hugh Dowding's Fighter Command in fending off the attacks of Goering's Luftwaffe and preventing a German invasion in the summer and autumn of 1940. The first RAF heroes were not the bomber crews but the dashing fighter pilots who streaked across the sky in their Spitfires and Hurricanes.

Churchill instinctively recognised that the role of the bomber was to take the offensive action that he so dearly wanted to pursue. In early September 1940, while the Battle of the Britain was still raging above southern England, he wrote in a directive to ministers:

The Navy can lose us the war, but only the Air Force can win it . . . The Fighters are our salvation, but the Bombers alone provide the means of victory. We must therefore develop the power to carry an ever-increasing volume of explosives to Germany, so as to pulverise the entire industry and scientific structure on which the war effort and economic life of the enemy depend.[3]

But Churchill's ambitions for RAF Bomber Command would be totally impossible to achieve for some time to come.

For many months, photographic reconnaissance of sites
that had been bombed continued to show that the damage
caused was negligible. This was not what the chiefs of Bomber
Command wanted to hear and they did not believe it could
be true. They came up with a variety of reasons to disprove
the photographic evidence. The photographs were of too small
a scale to be able to spot the damage. The photo interpreters did
not know what they were looking for. Some damage assessments came back with a note in the margin saying simply: 'I
do not accept this report.' But the morale of the bomber crews
who nightly put their lives at risk for what seemed no tangible
gain slowly deteriorated. Some parts of Bomber Command
did begin seriously to explore ways of improving navigational
techniques. But overall, a growing sense of disappointment per-
vaded the missions. No one was particularly to blame for this.
Years of financial cutbacks throughout the 1930s were once
again showing through, just as they had in the army and navy.
The RAF had poor aircraft. The Hampden could fly at only 155
m.p.h. The bombers could carry only small bomb loads that
were largely ineffective even if, by some near miracle, they hit
their target. And the crews lacked the navigational training
to fly across hundreds of miles of occupied territory at night,
often in freezing conditions and without properly pressurised
cabins. It was hopeless to expect much from them, so although
each night the BBC would broadcast that bombers were striking
such-and-such armaments factory, the horrible reality was that
it was mostly a waste of time. Frequently, from the dispersal of
the bombs dropped, German Intelligence could not only fail to
work out what the RAF's targets had been, but even which part
of Germany they had been trying to hit.

Lord Cherwell, as we have seen, was never far from Churchill's
side as his all-purpose scientific adviser. In late 1940, the Prof
began to raise doubts in Churchill's mind about the accuracy of

the RAF's bombing. The following year, Churchill instructed the Prof and a team from his statistical branch to carry out an investigation. David Butt of the War Cabinet secretariat analysed more than six hundred aerial photographs of post-bombing damage taken in June and July 1941. This analysis confirmed Cherwell's worst fears. When the moon was full, only 40 per cent of planes dropped their bombs within five miles of the target. In the absence of moonlight, a mere 7 per cent, only one out of every fifteen aircraft, got their bombs within five miles of the target. The Butt Report, as it was called, was immediately repudiated by the air marshals. Cherwell agreed that it was not a strictly accurate guide, but he argued that the figures 'are sufficiently striking to emphasise the supreme importance of improving our navigational methods'.[4] Churchill later wrote: 'The air photographs showed how little damage was being done. It also appeared that the crews knew this, and were discouraged by the poor results of so much hazard. Unless we could improve on this there did not seem much use in continuing night bombing.' He forwarded the report to the new Chief of the Air Staff, Air Chief Marshal Sir Charles Portal, with a covering note: 'This is a very serious paper, and seems to require your most urgent attention. I await your proposals for action.'[5] The future of the whole bombing campaign was in the balance.

In truth there was nothing much that Bomber Command could do at this point in time. The technology for improved navigational aids still lay in the future. The only realistic response was to abandon the concept of night-time precision bombing of specific factories or railway yards and to opt instead for the wholesale smashing of German cities on an indiscriminate basis, known as 'area bombing' or more collo-quially as 'carpet bombing'. In September, Portal told the Prime Minister that if he were given a force of four thousand heavy bombers, he could bomb different cities each night and 'break Germany in six months'. This was dangerously close to the wild claims made by Goering in the summer and autumn of 1940 about what his Luftwaffe could do to Britain.

Churchill was in a dilemma. He had largely lost confidence in bombing as a key strategic offensive weapon against Germany. He wrote to Portal: 'It is very disputable whether bombing by itself will be a decisive factor in the present war. On the contrary, all that we have learnt since the war began shows that its effects, both physical and moral, are greatly exaggerated.'[6] After all, it was known that the Blitz had failed to destroy Britain's war economy and in the long run it had only increased the resolve of the British people to fight on. On the other hand, with the Soviet Union now in the war fighting for its survival, it was clear that one of the few ways in which Britain could help its new ally was by striking back through a bombing offensive to undermine the German war machine. The Chiefs of Staff, with Churchill's backing, had repeatedly stated that this was a major objective of the British war effort. Could it be abandoned now? And who would explain such a reversal of policy to Stalin?

The debate between Churchill and Portal continued. Then, on the night of 7 November 1941, four hundred aircraft were sent by Bomber Command to attack Berlin, the Ruhr and Cologne. Thirty-seven of the bombers, nearly one-tenth of the force, were shot down or failed to return. This was an unsustainable level of loss. Air Marshal Sir Richard Peirse, the commander of Bomber Command, was a guest at Chequers the following night. Churchill told Peirse how worried he was by the rate of casualties. He went further and said to Peirse that 'he did not think we had done any damage to the enemy lately'. Churchill insisted there must be a break in the bombing offensive in order for Bomber Command to 're-gather their strength for the spring'.[7] This matter was discussed by the War Cabinet and it was agreed that it was pointless to fritter away the bombing fleet in a series of small and largely ineffectual raids. It now became official policy to rest Bomber Command through the winter months. This was a vote of no-confidence in bombing as a strategy to win the war, as well as a vote of no-confidence in Peirse, who was soon removed from his command and sent to the Far East. Churchill needed a new, more dynamic figure to take over Bomber Command and realise the potential of the bombing campaign against Germany.

Arthur Harris was appointed as the new head of Bomber Command in February 1942. Harris shared many of Churchill's traits. He wanted to take the war to the enemy. He believed totally that you had to take offensive action to win wars. Harris was also a disciple of Trenchard and his belief in the power of bombing. He wanted to bomb German cities as powerfully and as destructively as the strength of his force would allow. This was how to win the war, he believed, hence his nickname, 'Bomber' Harris, although his crews knew him as the 'Butcher', usually abbreviated to 'Butch'. He was single-minded about this task – to the point of alienating other senior RAF figures who wanted him to take a more all-round view of the war. He opposed sending his valuable heavy bombers on long-range patrols to hunt down U-boats on the grounds that these were distractions to the main task. (He did let up slightly when it was pointed out to him that if Britain did not win the Battle of the Atlantic, there would be no petrol for his bombers.) Churchill liked his commitment and his realisation that tough things had to be done to win the war. Harris has been a controversial figure ever since, but he was a resolute commander who did not flinch when his own losses mounted horribly, or at the thought of the destruction his crews rained down on German cities. The nature of his character is revealed by a story of his pleasure in driving his two-seater Bentley fast on the roads into and out of London. One evening he was stopped by a traffic policeman who said reproachfully, 'You might have killed someone, sir.' Harris replied sombrely, 'Young man, I kill thousands of people every night.' He seemed to revel in the hard-man role he had cast for himself.[8]

Bomber Command headquarters, where Harris worked, was at High Wycombe in Buckinghamshire, not far from Chequers. In March 1942, Harris paid his first visit to the Prime Minister's country residence and spent an evening with Churchill. The two got on well, and Harris would have many more invites over the next two years, sometimes for an evening of chat, sometimes for a weekend as part of a larger company. This direct access to Churchill's ear was an important element in Harris's growing stature. And it generated much resentment, particularly among the admirals, who thought that Harris was advancing the cause of his command at their expense. But Churchill had confidence in Harris, and Harris took inspiration from their regular conversations. He later wrote: 'The worse the state of the war was, the greater was the support, enthusiasm, encouragement and constructive criticism from this extraordinary man . . . He did not mind your expressing views contrary to his own but he was difficult to argue with for the simple reason that he seldom seemed to listen long to sides of the question other than his own.'[9] Harris soon discovered, as others had done before, that the best way to get a new point across was to write a short two- or three-page note on a subject. The rapport Harris built up with Churchill was based on mutual respect rather than on real friendship, and it waned towards the end of the war. But like General Brooke as CIGS, Harris would stay in his post until final victory, and he was another core member of Churchill's War Lab.

In his first few months as head of Bomber Command, Harris
orchestrated a few stunts to publicise and promote his new role.
In April, a force of low-flying heavy bombers struck deep inside
Germany at the U-boat engine production plant in Augsburg.
For his daring and courage, the leader of the raid, Squadron
Leader J.D. Nettleton, won the Victoria Cross. Then, after get-
ting Churchill's agreement at a nocturnal session at Chequers,
Harris launched the first 'thousand-bomber' raid of the war on
the night of 30 May 1942 against the city of Cologne. This was
lauded by the press and the newsreels, who were delighted that
Britain was hitting back hard. Headlines talked of 'The Biggest
Bombing Raid In History'. Both the Americans and the
Russians were impressed. In fact, Harris had managed to rustle
up the magic number of a thousand bombers only by deploy-
ing all his reserves, including every available training plane.
Instructors as well as trainees flew on the raid. It was a huge
risk but it paid off. Morale in Bomber Command shot up
overnight. The damage to Cologne, though less than was
claimed, was considerable. However, two later huge raids
proved far less effective. A thousand bombers tried to hit the
Krupp armaments factories at Essen in the Ruhr in June. But the
target was more difficult to find and the bombs were widely
dispersed. And nine hundred bombers targeted Bremen at the
end of June, inflicting severe damage on the port but suffering
substantial losses, too. Harris had put Bomber Command on
the map, but raids on this scale could not be sustained and he
had to settle for far more modest sorties for some time to come.

In early 1942, just as Harris was taking over at Bomber
Command, there was an intensification of the debate about the
bombing offensive. Cherwell once again took the initiative and
instituted an investigation led by two scientists, Solly Zuckerman and J.D. Bernal. Zuckerman, as we have seen, had already
carried out research into the impact of bombs on human beings

and on their physical surroundings. Now, he and Bernal studied the German bombing of Birmingham and Hull with a view to finding some general principles for the impact of a bombing campaign upon urban centres. Cherwell kept in close touch with Zuckerman and Bernal, but before they had completed their research, he wrote a paper which he sent to Churchill on 30 March. In this he claimed that one ton of bombs dropped on a built-up area demolished between twenty and forty houses and turned '100–200 people out of house and home'. So if each of Britain's new heavy bombers were to drop about forty tons of bombs during its service life, then each would make between four thousand and eight thousand people homeless. He then calculated that if only half of the bombers in a force of ten thousand aircraft dropped their bombs on the biggest cities in the German Reich, 'about one third of the German population' would be made homeless. He concluded that the area bombing of German cities could therefore 'break the spirit of the people' and prompt a breakdown of communications and supply lines and the collapse of public services.[10] In other words, bombing could definitely help the Allies win the war against Germany.

Cherwell's paper is often said to be the prime document that persuaded Churchill to restart the bombing offensive. In fact, it provoked great controversy at the time and has continued to do so ever since. The debate about the impact of bombing opened up the old rivalry between Cherwell and Tizard, whose disagreement had first erupted nearly two years before, in the summer of 1940. Tizard criticised the mathematics by which Cherwell had made his calculations and challenged his assumption that there would ever be a force of ten thousand heavy bombers. He told Cherwell: 'I am afraid that I think the way you put the facts as they appear to you is extremely misleading and may lead to entirely wrong decisions being reached with a consequent disastrous effect on the war. I think, too, that you have got your facts wrong.' Professor Blackett, the doyen of Operational Research, also weighed into the debate. He said he thought Cherwell's estimate of what could be achieved was 'at least six hundred percent too high'.[11]

Cherwell wasn't the type to take a rebuke like this sitting down. He had Churchill's ear and he continued to press his case, claiming that his figures had merely been presented in a way that meant Churchill did not need to do the mathematical calculations himself. As we have seen, this was typical of the way Cherwell presented his papers to a Prime Minister who found mathematical explanations too complex to follow. In the end, of course, it was Churchill who had to take the final decision when it came to resolving the argument for competing claims to limited resources. And he took the side of Cherwell and Harris, against Tizard and Blackett. The official historians of the bombing offensive concluded that, because of his position when he submitted this note, 'Cherwell's intervention was of great importance. It did much to insure the concept of strategic bombing in its hour of crisis.'[12] The bombing offensive against Germany was on again. Area bombing was the new policy, and this time Bomber Command had a combative leader with great ambitions for his growing fleet of bombers.

There is no doubt that part of what lay behind Churchill's decision was his desire to show Stalin and the Soviet people that, while there was no second front in Europe, at least Britain was striking hard at the heart of the Nazi war machine. Accordingly, Churchill decided to make a personal visit to Stalin and to explain this face to face to the Soviet war leader. After his momentous visit to Cairo in August 1942 when he set the 8th Army on the path to victory (see Chapter 6), Churchill took an overnight flight in his uncomfortable Liberator bomber to Teheran. From there, he took another long, gruelling flight to

Moscow, skirting the fighting that was raging below. He was accompanied by Averill Harriman, representing the US President, and the British Chiefs of Staff. The Soviets had been pleading for a second front to be opened in France, so Churchill's principal mission was to explain why this could not be done, and to sell Operation Torch and the importance of the renewed bombing campaign to the Russian leader. It was, as Churchill confessed to Roosevelt, 'a somewhat raw job', but he wanted to express his personal support by taking the long, arduous route to Moscow in order to explain exactly what Stalin's allies were doing to help him in the gargantuan struggle playing out on the Eastern Front.

A few hours after his arrival in Moscow, on the evening of 12 August, Churchill had his first meeting with Stalin in the Kremlin. It lasted four hours. Stalin told Churchill about the immensity of the struggle the Red Army were engaged in along the Eastern Front. He said German troops were now beginning to press on the industrial city of Stalingrad on the river Volga. For his part, Churchill wanted to get all the bad news about not opening a second front over with first. As he listened, Stalin became glum and grew restless. He cheered up when he heard about the plans for the Torch landings in North Africa. When they went on to discuss the bombing offensive, Stalin said that he attached the greatest importance to this strategy and that he knew the air raids were having a tremendous effect upon morale in Germany. Churchill promised he would press on with the raids and would 'show no mercy' to the German people. Stalin smiled and said, 'May God prosper this undertaking.'[13] So far, so good.

The next meeting did not go well. It started at 11 p.m. in the Kremlin. Stalin accused Churchill and the British Army of being frightened of the Germans, and claimed that Britain and the United States had failed to deliver the supplies promised to

Russia, keeping the best for themselves. Then he accused Britain and the United States of reneging on firm promises to launch the second front in Europe. Churchill stood his ground and disagreed point by point. The dictator was not used to being contradicted. At one point, Churchill raised his voice and said in a passionate outburst that he had come a long way to establish good relations with Stalin and that victory must not be undermined by disagreements that could benefit only the enemy. Before this could be translated, Stalin responded by saying that he liked the tone of Churchill's speech. But it was clear that Stalin's position had hardened, and this did not look good for Anglo-Soviet relations.

The following night, Stalin hosted a dinner for his guests. A lavish banquet was served and several toasts were washed down with vodka. Churchill sat on Stalin's right, and although no serious business was done, the Soviet leader seemed in a far more friendly mood. The issue of Churchill's hostility to the birth of the Soviet Union came up and Churchill openly admitted this had been the case. Stalin smiled and Churchill asked, 'Have you forgiven me?' Stalin replied, 'All that is in the past and the past belongs to God.'[14] When Churchill left, exhausted, at 1.30 p.m. (Stalin's late-night hours even outdid his own), Stalin insisted on escorting him through endless corridors and staircases to the Kremlin's front door. The British Ambassador had to trot to keep up with the two leaders, and he later said that he had never known Stalin do this for any other guest.

During the next day, General Brooke, Air Marshal Arthur Tedder and General Wavell (who was there because he spoke fluent Russian) had detailed talks with their Soviet military counterparts. The discussions did not go well. When the British delegation asked for information, they were repeatedly told that the Soviet chiefs had 'no authority' to release such details. Instead, they kept repeating that they wanted a second front *now*. After the openness and informality of relations with their American allies, these talks were a very different affair. Brooke was particularly annoyed by the Russians and brought the series of meetings to an abrupt conclusion.

On the final evening, Churchill went once again to the Kremlin for a last hour with Stalin at 7 p.m. As Churchill got up to leave, Stalin became very chummy and invited him back to his private apartment for some drinks. There, in the Soviet leader's private dining room, his aged housekeeper began to assemble a meal. Stalin's pretty young daughter appeared and laid the table while her father uncorked an impressive array of bottles. With just Stalin, Molotov (the Soviet Foreign Minister) and Churchill present, each with his own translator, the two war leaders and one time political foes talked on into the night. Occasionally they argued but always in a good-humoured way. No doubt the quality of the wine helped. Churchill talked about one of his pet projects, to invade northern Norway in order to create an easier passage for the Arctic convoys. They discussed the collectivisation of Soviet farming that Stalin had forced through in the pre-war years at the cost of millions of lives. At about 1 a.m., a huge suckling pig was brought in. Churchill finally left at about two-thirty. He felt that his mission to Moscow had been accomplished. He had built up a rapport with the Soviet leader.

As Churchill departed by air at dawn, only a couple of hours later, he was exhausted and, unusually for him, he had a splitting headache. Over the next few days, he had time to reflect on his meetings in Moscow. He had been seriously offended by some of what Stalin had said, especially about Britain's cowardice in the face of the Nazis. And he was genuinely puzzled by the fluctuations in Stalin's mood. On the other hand, he appreciated that, with the Wehrmacht only fifty miles from Moscow and the huge struggle with Nazi Germany nearing its climax, the news he had brought had been a bitter blow to the Soviet leader. More than anything, Churchill now realised that the Anglo-American leadership must push their war machines and their men to the maximum in their campaigns in North Africa and in the bombing of Germany. The bombing offensive now took on a political role as well as being a military tool. Everything that could be done must be done to help their Soviet allies defeat the Nazi invaders.

So Churchill returned to Britain with renewed enthusiasm for the bombing offensive. Within weeks, he had agreed to increase Bomber Command from thirty-two to fifty operational squadrons. He gave support to Harris in his internal struggles within the RAF. But once again, conflicting priorities held back any dramatic developments and no further thousand-bomber raids took place that year. The war against the U-boats continued to raise questions about the deployment of long-range bombers over the Atlantic. And the war in North Africa dominated Churchill's and the nation's attention. Furthermore, the promised arrival of large numbers of US bombers failed to materialise during 1942. All the bombing raids on Germany that Churchill had eagerly boasted about to Stalin were carried out by the RAF with their meagre, but growing, resources.

Most importantly for the future of the bombing campaign, a series of significant developments took place in 1942. First, there were major advances in technology that enabled RAF bombers to find their targets with far greater accuracy. In an attack upon the Renault factory near Paris, in March 1942, a new navigational aid called 'Gee' was used for the first time. Gee had been developed by scientists at the Telecommunication Research Establishment (TRE) near Swanage in Dorset. Every Sunday, open meetings were held at TRE between operational flying officers and the scientists, and everyone, regardless of rank or status, was encouraged to speak their mind. They were known as the 'Sunday Soviets'. At one of these meetings in June 1940, a senior staff officer had deplored the inadequate results of the RAF's bombing missions. In response, one of the scientists came up with an old notion that had been partly developed before the war for sending out a set of pulses as a blind-landing aid. A group set to work to improve this system and eventually came up with Gee. A series of pulses was sent from three transmitting towers in southern Britain. As the signals from the first tower hit those from the other two, they created a lattice-type network extending for a few hundred miles across Northern Europe. A navigator on board his aircraft could then pick up these signals with a cathode-ray tube and read where he was on a chart that marked the grid of beams across Europe. Gee was accurate to within two miles. This was an enormous advance over navigating by identifying landmarks below or taking readings from the stars above. Gee was also better than the German *Knickebein* system and its follow-ups because the British grid covered the whole of Europe and the beams were not intended to guide aircraft to one particular spot. But the RAF was worried about one of its planes coming down over occupied Europe and the cathode-ray tube and accompanying charts falling into enemy hands. The Germans might then be able to figure out how the system worked and could even try to distort the pulses. So an elaborate deception was set up whereby it was leaked to the Germans through double agents working in Britain that the RAF now had a 'J' system (which sounded close to 'G'), based on the German *Knickebein* beams. A set of meaningless beams was duly transmitted and German Intelligence spent some time tracking them down and twisting them, while failing to identify the pulses that made up the new Gee system. Gee was successfully used for some years and remained in use after the war with the RAF and as an aid for shipping.

In addition to the development of this navigational aid, the RAF also needed a more precise blind-bombing guide, which would enable aircraft not just to bomb one part of a city, but to find and bomb a particular factory or military installation. Once again, it was the scientists at TRE who came up with the solution. It was called 'Oboe'. This system was based on two radar transmitters, one in Norfolk, known as 'Cat', and one in Kent, known as 'Mouse'. The two stations were linked and transmitted synchronised pulses. The pilot flew along the Cat beam across Europe towards the target. The operators of the Mouse beam could follow the aircraft and calculate its precise location. When it was over the target, they would transmit a signal to the plane's navigator, who would drop the bombs. Oboe had an accuracy of about a hundred yards, which was of a different order to anything that had gone before. When Oboe became operational in 1943, it was used by accurate Pathfinder aircraft, like the super-fast de Havilland Mosquito. These aircraft identified and hit the target with flares or incendiaries that the other bombers would then use as location finders.

Both Gee and Oboe were real achievements for the scientists, more strikes for Churchill's wizards. But they had their limitations. Gee had a range of about 450 miles from the British coast and Oboe about 250 miles. This was good enough to hit the Ruhr and many other prime targets, but not enough to penetrate the whole of occupied Europe. Something better was still needed.

Despite his call for improved navigational aids, Cherwell himself was not involved with the development of Gee and Oboe. He was however closely involved with the development of a new radar mapping system code-named 'H2S'. This was carried inside aircraft and at first utilised the revolutionary cavity magnetron, which enabled radars to work on a wavelength of 10cm, rather than 150cm as before. Aircraft that were

Churchill and President Roosevelt at their first wartime meeting, August 1941, singing hymns together with their military chiefs. Churchill had specially selected the hymns to emphasise unity between the two nations. (IWM)

General Sir Alan Brooke, Chief of the Imperial General Staff from 1941–5. A tough Ulsterman, he met with Churchill almost daily and had an often stormy, love-hate relationship with the PM. (IWM)

Air Marshal Sir Arthur 'Bomber' Harris looks at aerial photos on the wall. Photo-interpreters study aerial images in the foreground. Harris was totally committed to bombing as the way to end the war. (IWM)

Churchill surrounded by military chiefs in North Africa, June 1943. From left to right, General Sir Alan Brooke (with papers on his lap), Air Vice Marshal Sir Arthur Tedder, Admiral Sir Andrew Cunningham, General Sir Harold Alexander, American Generals George C. Marshall and Dwight D. Eisenhower (sitting) and General Sir Bernard Montgomery (standing). (Getty Images)

Churchill in his favourite dressing gown patterned with red dragons, Tunisia, Christmas Day, 1943, with General Eisenhower who had just been appointed to command Operation Overlord. General Alexander behind them. Churchill was recovering from a serious illness that at one point threatened his life. (IWM)

Stalin proposes a toast to Churchill on his sixty-ninth birthday during the Teheran conference, 30 November 1943. Anthony Eden is on Churchill's right. The toasts went on throughout the evening. (IWM)

Churchill in Italy, August 1944, watching an artillery barrage on an enemy position. He and the generals were within enemy artillery range. On his visits, Churchill liked to get as near to the front as possible. (IWM)

Churchill arrives in Normandy six days after D-Day for a visit and is escorted off the beach by General Montgomery. Churchill had wanted to be present at D-Day itself. (IWM)

The British Mulberry harbour at Arromanches, June 1944. The huge caissons can clearly be seen along with miles of floating roadways. Churchill knew the building of two Mulberries would be a gigantic undertaking but persisted. Fittingly this one became known as Port Winston. (Courtesy of the Medmenham Collection)

A chilly day at the Yalta Conference, February 1945. Behind the Big Three stand Anthony Eden, Foreign Secretary; Edward Stettinius, Secretary of State; and Vyacheslav Molotov, Soviet Foreign Secretary. Roosevelt is visibly ailing. He will not live to see the victory. (Corbis)

Churchill the conquering hero. With trademark cigar in mouth, he leads
Generals Brooke and Montgomery and a bevy of American starred
generals on to the east bank of the Rhine, 25 March 1945. Later that day
they came under fire and the PM had to be dragged away to safety. (IWM)

equipped with it could pick up buildings and the general landscape much more clearly. A rotating radar scanner was fitted in
a cupola below the belly of a heavy bomber and the navigator
read the landscape below on a circular screen. First trials took
place at the end of 1941. The team trying to develop this radar
mapping system encountered immense difficulties, including
the crash of the prototype bomber in which it was installed.
Many of the leaders of the research team were killed in that
accident. More delays followed when the RAF chiefs decided
that the cavity magnetron was so secret that it must not
fall into the hands of the enemy and therefore could not be
flown over occupied territory in case the plane crashed and it
was recovered by German scientists. So an alternative device,
the Klystron, was developed instead that was closer to known
existing German technology. The debate within the Air Staff
continued as to whether the concept of radar mapping could
ever work effectively, but Cherwell continued to support it and,
on his advice, so did Churchill. In June 1942, Churchill issued
one of his typical instructions to the Secretary for the Air, calling
for far greater effort in the manufacture of the new equipment
and the training of crews in its use. He insisted that 'nothing
should be allowed to stand in the way of this'.[15] The many
doubters would almost certainly have killed off the development of the ground-mapping radar system without Cherwell's
enthusiasm and Churchill's support. It finally came into operational use in 1943 and refinements continued to be made
throughout the war.

During 1942, another key development took place, the
arrival of the first of a new generation of heavy bombers.
Bomber Command had begun the war with their two-engined
Whitleys and Hampdens, later joined by the more robust
Wellington. But bombing on the scale that was now needed
required far larger aircraft that could carry much heavier bomb

loads. Britain's pre-war aircraft industry was lucky to have some visionary designers like Reginald Mitchell, who had designed the superlative Spitfire, and Roy Chadwick, who worked for the Avro company. Chadwick had already designed the two-engined Manchester bomber but decided it was too small to fit the bill, so he set to work on a four-engined giant that could carry more than three times the bomb load of the conventional bombers of the day up to a distance of two thousand miles. He used four Rolls-Royce Merlin engines, the same brilliant engine that was used in the Spitfire, to power his new beast, which was called the Lancaster. These new heavy bombers were designed to be mass produced and in early 1942 they began to come off the Avro production line, but still only in small numbers. It was not until the end of the year that the Lancaster and the other four-engined heavies, the Stirling and the Halifax, were available in significant numbers. The arrival of these heavies was similar in its impact to the arrival of jumbo jets to a later generation. Their vast scale seemed to dwarf almost everything that had flown before. Five thousand Lancasters would be built and the aircraft would come to symbolise the British bombing offensive against Germany, just as Boeing's B-17 'Flying Fortress' came to represent the American bombing campaign.

At the Casablanca Conference in January 1943, Roosevelt and Churchill fully endorsed the bombing strategy. Their directive to the British and American bomber commanders, which was approved by the Combined Chiefs of Staff, stated: 'Your primary objective will be the progressive destruction of the German military industrial and economic system, and the undermining of the morale of the German people to a point where their armed resistance is fatally weakened.'[16] A list of targets then followed, including submarine yards, aircraft factories, oil refineries, rubber and tyre plants, and factories for military transport. With American bombers now arriving in England in large numbers at last, and with all the improvements in technology and design, the bombing offensive would take off again during 1943. Churchill was keen to keep Stalin well informed of progress. He regularly sent him books of aerial photos of bomb-damage assessments. When congratulating the Soviet leader on his victory at Stalingrad, Churchill added, almost as a postscript: 'We last night dropped 142 tons of high explosives and 218 tons of incendiaries on Berlin.' Stalin replied: 'I wish the British Air Force further successes, most particularly in bombing Berlin.'[17]

The bombing offensive that now started in earnest against Germany was based on a strategy of RAF Bomber Command bombing by night, and the US Army Air Force (US AAF) bombing by day. The Americans had the Flying Fortress, which, with eight mounted heavy machine guns in four turrets and armour plating in the fuselage, was regarded as powerful enough to withstand attacks from enemy fighters. And the Americans had a bomb sight known as the 'Norden' that they regarded as supremely precise. 'You can drop a bomb in a pickle barrel from ten thousand feet' was the popular boast about the Norden. But it needed to be used in clear skies in daylight. The Americans rejected the policy of area bombing, which they thought was ineffective. Instead, they wanted to go for planned, precise attacks upon key elements of the German war economy that would cause maximum disruption. Churchill initially thought this was impossible to achieve, and along with Harris and his Chiefs of Staff wanted the Americans to join the night-time bombing campaign, adding their vast scale to the efforts of the RAF. But at Casablanca, Churchill met with General Ira Eaker, the commander of the USAAF in Britain. Eaker persuaded Churchill that by bombing around the clock, the combined RAF–USAAF operation would 'give the devils [i.e. the Germans] no rest'.[18] And so the strategy of

twenty-four-hour bombing was endorsed, with the Americans
going for precision bombing by day and the RAF opting for
area bombing by night.

Harris could now at last take advantage of the advances in
technology and make use of the growing supply of Lancasters.
As Max Hastings has written, only in early 1943 did all the
latest radio and radar equipment and the supply of heavy
bombers enable Harris 'to bring the bomber offensive out of its
cottage industry phase into the age of automated mass destruction'.[19] Most of the older, obsolete aircraft were pensioned off
and his main force of about five hundred bombers was in a
far better state to strike at the enemy than a year before. The
spring offensive began on 5 March with an assault upon one of
Harris's favourite targets, the giant Krupp armaments factory
at Essen. Mosquitoes equipped with Oboe acted as Pathfinders
to locate the target. The raid was a success, with a third of the
force dropping its bombs within three miles of the aiming
point, and 160 acres of Essen were flattened. But the Ruhr,
Germany's industrial heartland, was well defended. As more
raids on 'Happy Valley', as the bomber crews sarcastically
called it, followed over the next three months, the number of
bombers being shot down by radar-assisted German night
fighters grew to alarming levels. Then, in July, Harris tried
something new.

The scientists had come up with another simple innovation,
the dropping of thousands of small metal foil strips which
totally blinded the enemy radar. It was known as 'Window'.
The technique had been worked out earlier, but Cherwell had
pointed out that, as there was no known remedy for this, if the
RAF tried out Window over Germany, the Luftwaffe was
bound to use it when it next bombed Britain. Its use was
delayed, and again it needed Churchill to intervene to settle
the dispute. 'Let us open the window!' he ordered at the end

of June, and on 24 July Window was used for the first time in a heavy raid on Hamburg. It was a total success. The night fighters completely failed to find the bombers and the mission was carried out with copybook precision. The US bombers followed up this initial attack with two days of daylight bombing on Hamburg. Then the RAF returned on the 27th and dropped incendiaries on the city. Fires were started that raged for days, strengthened by two further raids by Bomber Command over the next week. The devastation was extreme. The fires reached temperatures of a thousand degrees centigrade, and as the heat rose it sucked in more oxygen, creating whirlwinds or hurricanes of flame in a deadly, blazing inferno. Buildings disappeared in the firestorm. Bodies were incinerated. When the fires abated, it was estimated that 42,000 civilians had been killed, a million had fled into surrounding areas, and twenty-two square kilometres of the city had been razed to the ground. Forty thousand houses and more than five hundred factories had ceased to exist. The chief of the Hamburg fire brigade captured the terrible inferno on colour film which provides a permanent record of the terror of the fire raids.[20]

The Nazi leadership was truly shaken for the first time. Goebbels confided in his diary that the bombing of Hamburg was a 'catastrophe, the extent of which simply staggers the imagination'. Albert Speer, who had become Minister for War Production in February 1942, later said he told Hitler that if the Allies had mounted similar attacks against six cities in quick succession, German war production might have collapsed.[21] For Harris, of course, this was a triumphant success, but Bomber Command was wary about returning to the same location night after night, in case enemy fighters were waiting, and the terrible carnage of Hamburg was not replayed on other German cities until towards the very end of the war.

As the American fleet of heavy bombers grew in Britain, the debate about the objectives of the bombing campaign continued. After a discussion in Washington which Churchill, unusually for him, nodded through without challenge, the Combined Chiefs of Staff agreed to pursue Operation Pointblank. This summarised the objectives of the bombing and clearly shows that American influence dominated. A priority was given to targets relating to the German aircraft industry, with both factories and airbases to be attacked to eliminate the threat from the German fighters. Other industrial targets were also listed, and Pointblank referred to the need to try to destroy German morale. Harris seized on this last point to justify the continuation of area bombing, which he was convinced would win the war. On the other hand, the American commanders, Hap Arnold in Washington and General Carl Spaatz, head of the 8th Air Force in Britain, used Pointblank to justify precision attacks on industrial targets. From the start, the two bombing efforts increasingly went their separate ways.

Through the autumn and winter of 1943, Bomber Command turned its focus to what became known as the 'Battle of Berlin'. Twenty thousand sorties were flown against the German capital. Harris was supremely confident that this was the best way to defeat the Nazi war effort. But the city was out of range of Oboe so the crews relied upon H2S radar to identify their targets, a far more difficult tool to use effectively. Air defences grew in scale and expertise around the city. Huge towers were built to house powerful 88mm anti-aircraft guns. Losses mounted. Harris predicted to Churchill in November that 'We can wreck Berlin from end to end if the USAAF will come in on it. It will cost [us] between 400–500 aircraft. It will cost Germany the war.'[22] This was a rash claim that was used later against Harris. But by now Churchill did not interfere much in the development of air policy. He left it to Harris and his team to select targets and to pursue their aims. Harris became an increasingly rare visitor to Chequers. However, in March 1944, RAF losses from night-time bombing of Germany rose again to unacceptable levels of about 5 per cent of total aircraft. On one raid on Nuremberg it was as high as 9 per cent. With these levels, crews could not expect, statistically, to survive a single tour of duty. A pause was called in the bombing offensive. By now, fortunately for the crews, other priorities had arisen.

American confidence in the firepower of their Flying Fortress had proved not to be justified. Their daylight raids, flown unescorted over Germany, were subject to intense harassment from German fighters. Large numbers of B-17s were shot down and each aircraft down meant the loss of the ten members of its crew. Following their policy of hitting key targets, the US 8th Army Air Force based in Britain flew a series of raids in October 1943, culminating in an attack on the ball-bearing factory in Schweinfurt, deep in the heart of Germany. On the night of 14 October, 60 aircraft out of 291 on the raid were shot down. During that week, the Americans lost a total of 148 bombers. The US bombing offensive against Germany was postponed until long-range fighters were available to escort the bombers. This finally came about in the spring of 1944, when the P-51 Mustang was fitted with long-range fuel tanks, enabling it to fly all the way to Berlin and back. The sleek, silver Mustangs could defend the bombers from attack by enemy fighters, and on their return were told to seek out 'targets of opportunity', which usually meant swooping down to attack enemy air-craft on the ground. Only now did the Allies win daylight air supremacy over Germany. The story goes that when Goering looked up and saw enemy fighters over the skies of the German capital he pronounced: 'The jig is up,' meaning the war was lost.

The RAF had not entirely given up on daylight precision raids, although Harris was bitterly opposed to them as they distracted from the heavy area bombing that he favoured. Tension grew between the Ministry of Economic Warfare, which wanted to set targets whose destruction it believed would be critical to German's war economy, and Harris, who demanded the right to select his own targets. The ministry identified the dams that supplied the water and produced much of the electricity for the Ruhr as a key economic target. In early 1943, the inventor Barnes Wallis developed a way of spinning cylindrical, depth-charge-type bombs that could be dropped by low-flying Lancasters as 'bouncing bombs'. The idea was to bounce them over the German defences and destroy the dams. Harris was initially hostile, describing bouncing bombs as 'the maddest proposition as a weapon that we have yet come across'. He allowed one aircraft to be used for experimental purposes, but was not prepared to hand over any more of his precious Lancasters.[23]

However, on meeting Barnes Wallis in person, Harris was persuaded to give the idea a go. He agreed to create a special team, 617 Squadron, to train in the use of the bouncing bombs, and he suggested putting Wing Commander Guy Gibson in charge. Gibson was allowed to select his own elite crews from across Bomber Command. The attacks took place on 16 May 1943 and involved low-level precision flying across Holland and Germany by nineteen Lancasters. The heroics of the Dam Busters' raid, in which the bouncing bombs were dropped from only sixty feet, are now well known mainly from the triumphalist feature film that became a war classic in the 1950s.[24] The Mohne Dam was successfully breached and the Eder was damaged. However, the critical third dam at Sorpe was left intact and the power supply to the industries of the Ruhr was maintained. Despite this, Albert Speer later wrote that if the attack had been followed up, it could have ended the war that year.[25] The raid was immensely popular with the British press and public, and Guy Gibson was awarded a Victoria Cross for his heroism in leading the mission. However, as eight of his best crews had been lost in the raid, a 40 per cent loss rate, Harris concluded that such missions were not worth the cost. The ultimate lesson he took from this daring raid was of the need to maintain instead the policy of area bombing.

An aircraft more suited to low-level precision flying was the twin-engined Mosquito. With a top speed of nearly 400 m.p.h., it could out-fly just about anything in the pre-jet era. The frame of the Mosquito was built out of wood to get around the shortage of metal in wartime Britain. It was produced in large numbers but was in heavy demand for a variety of roles: as a night fighter, for photo reconnaissance, as a Pathfinder and as a light bomber. The Mosquito was used for several extraordinary operations. The most daring of these was a raid on the prison in the northern French town of Amiens on 18 February 1944. British Intelligence wanted to rescue a group of French Resistance leaders who were about to be executed by the Gestapo. A squadron of Mosquitoes was tasked with knocking down the walls of the prison, but in such a way as not to harm most of the prisoners, who would then be able to escape. The mission was called 'Operation Jericho'. Such accuracy would be difficult to attain even in today's era of GPS-guided missiles, but the Mosquito squadron achieved its objective with a superb display of flying. The planes approached fast and low, at about fifty feet, and hit the prison walls with pinpoint accuracy. Some prisoners were inevitably killed but over 250 French prisoners escaped in the resulting confusion.

Churchill loved to hear of missions like this. They were the RAF equivalent of commando raids. Soon he became involved in one debate which did not attract much attention at the time, but has since taken on major significance. In the summer of 1944, more and more shocking stories leaked out of occupied Europe about a terrible extermination camp in Poland where hundreds of thousands of Jews were being sent. Some of the thousands arriving each day were picked out by the SS guards as fit and able and were assigned to a brutal work camp, where most did not survive for long. The majority were sent straight to gas chambers, where they were killed. Their bodies were then burned in giant crematoria. The whole camp had been laid out with chilling efficiency to process and kill on an industrial scale – up to twelve thousand people each day. Its name was Auschwitz.

On 6 July 1944, the Jewish Agency sent two senior representatives to meet Anthony Eden, the Foreign Secretary, in London. They reported a sudden increase in the number of people being taken to Auschwitz as the Nazis geared up to exterminate the Jews of Hungary. They calculated that one and a half million Jews had already been killed. During the meeting, they made a formal request to bomb the camp. Eden passed this request on to Churchill, who had been a supporter of the Zionist cause since his visit to Palestine in 1921. He was shocked by the appalling estimates of the number of deaths. Churchill responded the following day by saying he agreed with the request to bomb the camp and wrote: 'Get anything you can out of the Air Force, and invoke me if necessary.'[26] Eden duly passed the request on to Sir Archibald Sinclair, the Minister for the Air.

There were several obvious difficulties with such a proposal. First, of course, the precise location of the camp had to be identified. Then a way of bombing it that minimised the risks to the inmates had to be found. The RAF considered the proposition and also considered bombing the railway lines leading to the camp. They concluded that although precision raids had been carried out in France, an attack on such a small target in distant Poland was virtually impossible. If it could be done at all, it could be achieved only by transferring much sought-after Mosquitoes from other essential duties. It was a critical point in the war. The Allies had still not broken out from the bridgehead they had established in Normandy. Flying bombs were being fired on London every day and the effort to locate and destroy their launch sites was a major priority. Neither Churchill nor Eden pressed the case and the request to bomb the camp was dropped as impractical. In an ironic twist, in July 1944, aerial reconnaissance photos were taken of the Monowitz chemical plant in Poland. Just outside this plant was the giant complex of Auschwitz–Birkenau. The camp was caught in perfect detail on the aerial photos. On one set, a train has just arrived and the SS are in the process of separating those who are disembarking. But the photo interpreters were not looking for an extermination camp and they never identified Auschwitz as such at the time. So the Allies did have a record of the exact location and layout of Auschwitz – even the gas chambers and ovens can clearly be seen.

In recent debates about the Holocaust, there has been much criticism of the Allies for their failure to bomb the extermination camps. The reality is that even if they had analysed the pictures correctly, a bombing raid against Auschwitz would still have been an immensely difficult mission to carry out. Although in retrospect it seems criminally negligent that the Allies did nothing to prevent the mass murder of the Holocaust from continuing, at the time there were simply too many other military priorities in the war effort.[27]

In early 1944, all Allied planning became dominated by the preparations for the D-Day landings in Normandy, Operation Overlord. A new debate began as to what role the Allied bombers should play in this huge operation. Harris and his

American counterpart, Spaatz, persisted in their view that
bombing Germany was the main way in which the bombers
could contribute towards the success of D-Day, by undermining German war production and civilian morale. But there was
considerable scepticism among the army commanders about
the claims of the bomber barons. There was no sign yet of the
predicted collapse of the German war economy (as we shall see
later, output was actually increasing) or breakdown of support
for the war among the German people. General Eisenhower,
who was appointed Supreme Commander of the Allied Forces
for the invasion of France, insisted he must take direct command of the strategic bombing force. He wanted it to take on a
new role.

Solly Zuckerman, the scientist who had studied the impact of
German bombing on Birmingham and Hull, had gone on to
analyse the impact of Allied bombing on enemy targets in
North Africa, Sicily and southern Italy. This research had shown
the devastating military consequences of bombing the railway
networks. Zuckerman's work was enthusiastically taken up
by Air Marshal Sir Arthur Tedder, head of the RAF in the
Mediterranean, who was appointed Eisenhower's deputy.
Zuckerman and Tedder drew up a plan for the bombing of railway marshalling yards across northern France and Belgium so
that the Germans could not rush reserves to the area where the
landings had taken place. This became known as the 'Transport
Plan'. Eisenhower wanted to take command of the British and
American bombers based in Britain as a tactical bombing force.
Harris and Spaatz resisted this with all their energy. But Harris
had few friends left in the Air Ministry, where officials had
grown weary of his adversarial nature. He regarded any official
who was not 100 per cent behind him as someone who was
against him. It was reported that when he passed one particular civil servant in the Air Ministry he said: 'Good morning,

Abrahams, and what have you done to impede the war effort today?'[28] Neither was his cause helped by the inflated claims he repeatedly made for the success of his bombing campaign.

Churchill did not want to see the RAF fall under direct American command, and Cherwell was opposed to giving up the bombing offensive against Germany. It fell to Churchill himself eventually to come up with a compromise. Tedder was to have control of RAF's Bomber Command in the months before Overlord. This was acceptable to Eisenhower, and everyone was happy, except Harris. But Cherwell foresaw another problem. He calculated that bombing the French railways could kill up to forty thousand French civilians. Churchill took up this cause, worried that killing so many French men and women was not a good way to start a campaign to liberate their country. He wrote to Eisenhower in April that 'this might be held to be an act of very great severity' against Britain and America's 'friends'. The question was appealed to the President, who wrote back to Churchill: 'However regrettable the attendant loss of civilian lives is, I am not prepared to impose from this distance any restriction on military action by the responsible commanders that in their opinion might militate against the success of Overlord or cause additional loss of life to our Allied forces of invasion.' Roosevelt's response was, as Churchill later wrote, 'decisive'.[29] The Prime Minister raised no further objections. It was a sign of how far power and decision-making had shifted from the British to the Americans. Overlord was an American-led operation. The US commanders were calling the shots. And the President gave them his full support.

For several months, the heavy bombers targeted the railways of north-western Europe with great success. Coastal defences and other communication hubs were additional targets. Although as many as twelve thousand French and Belgian civilians were killed in the run-up to D-Day, this number was a lot less than had been feared. And the disruption caused inside France made it difficult for the German Army to bring up reserves in the wake of the invasion. The new tactical deployment of the bombing fleet made a major contribution to the success of this next vital chapter in the war.

Churchill's lack of interest in the bombing offensive in Germany from the summer of 1943 onwards is reflected in the lack of space he devotes to it in his war memoir–history.[30] This was partly down to his ambivalence about the role of bombing and partly down to the post-war debate about the wisdom and morality of the bombing campaign at the time when he was writing. But Churchill's influence was still to be felt over the final stages of the bombing offensive.

During 1944, the USAAF targeted German synthetic-oil production plants with stunning success and showed how effective the strategy called for in the Pointblank directive could be. The fuel supplies available to Hitler's army and air force were drastically reduced. The Allies now enjoyed total air superiority over Germany, with the surviving aircraft of the Luftwaffe largely grounded by their lack of fuel. Late in 1944, Bomber Command resumed its earlier policy of area bombing German cities by night. As the Red Army advanced westwards through Poland and approached the German border, there was a debate about how the bombers could be used in a way that would assist their progress. On 25 January 1945, a Joint Intelligence Committee report sent to Churchill and the Chiefs of Staff suggested using the heavy bombing force against targets south and east of Berlin. It was claimed this would disrupt troop reinforcements heading for the front, hamper the German administrative and military machine, and could even have 'a decisive effect on the length of the war'. Churchill discussed this with Archibald Sinclair and urged him to follow the recommendations and report back. Sinclair asked his staff to investigate the possibilities, explaining that Churchill was keen on 'blasting the Germans in their retreat from Breslau'.[31] This was reinforced when the Allied leaders met in early February at Yalta. Here the Soviet Chiefs of Staff made a formal request for British and American bombers to be used to paralyse German movements behind their retreating army.

The consequence of this was the combined bombing operation against the city of Dresden on the night of 13 February and over the following days. On the first night, a force of 900 RAF bombers dropped nearly 1500 tons of high explosive and over 1000 tons of incendiaries on the city. Hours later, US bombers followed this up with a massive daylight raid on the city. And that night the RAF returned. Once again, as in Hamburg, giant firestorms were ignited that destroyed vast areas of the city, leaving chaos and destruction in their wake. Dresden was an ancient city, a centre of art and culture, but much of its architecture was utterly destroyed in the flames. Days after the raid, photo interpreters were still unable to assess the damage because a pall of smoke hung across the city.

Dresden was packed with refugees fleeing the advancing Red Army, so it is impossible to know precisely how many people died. Estimates range from thirty thousand to about one hundred thousand. Whatever the true figure, the scale of destruction was immense and began to provoke revulsion. One news correspondent spoke of the 'deliberate terror bombing' of German cities. His report was censored in Britain but created a stir in the United States, where General Marshall made it clear that the raid was a consequence of a direct request from the Russians. Churchill had been involved in only the most generalised way by encouraging the implementation of an intelligence report to bomb eastern Germany. He had not mentioned Dresden specifically in conversations or in written minutes. But even he judged it was time to distance himself from the bombing campaign against Germany. In a memo on 28 March 1945, after he had been shown accounts of the raid, he wrote:

> It seems to me that the moment has come when the question of bombing of German cities simply for the sake of increasing the terror, though under other pretexts, should be reviewed ... The destruction of Dresden remains a serious query against the conduct of allied bombing . . . I feel the need for more precise concentration upon military objectives, such as oil and communications behind the immediate battle zone, rather than on mere acts of terror and wanton destruction, however impressive.[32]

The wheel had come full circle. Churchill, who in 1940 had believed that the RAF could win the war, and who had backed his scientific and military advisers in their arguments in favour of area bombing, was now veering away from the terrible consequences of this policy, calling for more precise targeting of military objectives. The time to reassess had come.

So was the bombing offensive worthwhile? Questions about both the effectiveness of bombing Germany and the morality of killing civilians have raged since the last Lancaster returned from Dresden. In all, about 600,000 men, women and children were killed in the bombing of Germany, with about 20 per cent of homes in German cities damaged by aerial bombardment. But the morale of the population did not break until the very last months of the war, and this was caused more by the realisation that defeat was imminent, with the Red Army advancing relentlessly into Germany from the east and Allied troops advancing rapidly from the west. Moreover, although there is still much debate about this, the German war economy actually increased its output during the years of the bombing offensive. Armament production increased by about 80 per cent in 1942; by another 20 per cent in 1943; and by a further 40 per cent in the first six months of 1944. In other words it roughly trebled over the two and a half years. But part of this is explained by the fact that the German economy was geared to fighting a short, successful war in 1940 and 1941 and was by no means fully mobilised in those years. There was much spare capacity to call upon. The Nazi creed believed, for instance, that women should be mothers and wives, not industrial labourers, in the early years of war. Women were not deployed at first to anything like the same extent as they were in Britain and would later be in the United States. Also, by the end of the war, the German war machine was able to draw upon the forced labour of millions of slaves who were brought to the Reich from occupied territories and put to work, often in appalling conditions. Albert Speer claimed that the indiscriminate bombing of vast tracts of German cities was not the best strategy for the Allied bombers. He insisted that if the Allies had pressed on with their bombing of key targets like the ball-bearing factories, or had mounted a series of devastating raids like the fire raids on Hamburg, then the bombing offensive would have had far greater impact. On the other hand, specific elements of the bombing campaign had very clear and damaging results. The attacks on oil refineries and synthetic-oil production plants in 1944 led to serious fuel shortages which undermined the German Army's ability to fight in the final campaigns of the war. Moreover, the bombing kept about one million soldiers at home to man anti-aircraft defences, equivalent to about fifty divisions which otherwise could have swayed the war on the Eastern Front or in Normandy. Another one and a half million workers were tied up clearing rubble and in reconstruction simply to allow life to go on.

Ultimately, it is an ethical decision as to whether the bombing campaign against Germany was justified or was a war crime. Certainly, Churchill was an advocate of the campaign, albeit an uncertain and questioning one at times. Cherwell and 'Bomber' Harris were its chief protagonists. It's interesting to note that in the moral climate of the war the argument against bombing was made not on ethical but on practical grounds. Was it worth the massive effort in industrial output and in the lives of the RAF crews? After all, as was said over and over again during the war, it was the Germans who had started bombing cities, and the memories of Warsaw, Rotterdam, and the Blitz on London and Coventry were strong in everyone's minds. Harris told the newsreel cameras, 'They who sow the wind will reap the whirlwind,' and most Britons at the time felt he was right. Churchill was encouraged to 'give one back' to Hitler. And, in purely numerical terms, how does one compare the 600,000 German civilians who died as a consequence of Allied bombs with the two million Germans who died at the hands of advancing Soviet soldiers at the end of the war? Or with the six million Jews who were murdered during the Holocaust? Or with the US bombing offensive in Japan? Of course, that campaign culminated in the dropping of atomic bombs on Hiroshima and Nagasaki, which prompted the final Japanese surrender and the end of the war.

The arguments will rage on. Churchill, as the final decision-maker in the British war machine, must take his share of the responsibility for the adoption of the policy of area bombing and its consequences. He definitely listened too closely to Cherwell and did not pay enough heed to other advisers who argued a contrary point of view. But this was all about practicalities and effectiveness, not about the morality of bombing. At the end of the war, when he was perhaps reflecting more on the verdict of history than he had done earlier, Churchill distanced himself from the bombing campaign. Air Chief Marshal Harris was left out of the principal victory celebrations and no campaign medal was ever given to the crews of Bomber Command, despite the courage they had shown and the losses they had endured – 57,000 airmen killed during the war. Even the erection of a statue to Harris in London fifty years after the end of the war created intense controversy. The bombing offensive against Germany without doubt remains the most heavily criticised element of British strategy during the Second World War. Any evaluation of Churchill's reputation as a war leader will forever be bound up with it.

9

Overlord

After the successful summit with Roosevelt at Casablanca in January 1943 (from the British perspective at any rate), Churchill flew to Turkey, where he tried to persuade President Inonu to join the Allied war effort, without success. He then went on to inspect a victory parade of the 8th Army in Tripoli. After all the upsets of the war in North Africa, Churchill found this celebration very emotional and tears ran down his face. He then flew on to Algeria before returning to London after another long flight in the primitive Liberator bomber. When he got back to Paddington Station, thirteen ministers headed by Attlee, Eden and Bevin turned out to welcome him on the platform. He had been away for twenty-six days. Within a week of his return, he had been struck down with pneumonia. He was largely out of action for another month but, despite the illness, he continued to dictate minutes. General Brooke, who had been travelling with Churchill, was himself struck down with influenza for two weeks. Both men's illnesses are a sign of the strain they were under during these intense meetings abroad and the long, gruelling and sometimes dangerous travel involved.

By mid-March, Churchill was fighting fit again and back in control. But he still had to wait for the breakthrough in North Africa. Hitler had decided to reinforce his troops in Tunisia, and the battle to evict the German and Italian armies was progressing far more slowly than planned. Rommel was fighting a superb campaign of withdrawal. The raw American troops suffered a severe setback at the Kasserine Pass in February, and it took several weeks for their advance to pick up momentum again. Churchill once more tried to whip on his commanders and sent off a missive complaining about the 'many factors of safety' that were creeping into operational plans, which meant they were 'ceasing to be capable of making any form of aggressive war'.[1] As always, he called for his soldiers, sailors and airmen to take an aggressive line. In March, Montgomery broke through the German defences known as the 'Mareth Line'. And, finally, the following month, the American armies advancing eastwards met up with the 8th Army advancing westwards. The Axis forces were now surrounded and the Allied armies edged forward towards the city of Tunis, the site of Rommel's last stand in North Africa.

In early May, Churchill once again felt the need to cross the Atlantic with his Chiefs of Staff for another Washington conference. Planning for the invasion of Sicily, agreed at Casablanca, had fallen behind. Eisenhower sent a note to the Combined Chiefs of Staff saying that if there were more than two German divisions in Sicily, the landings might have to be postponed. Churchill erupted. He wrote to his Chiefs of Staff: 'If the presence of two German divisions is held to be decisive against any operation of an offensive or amphibious character open to the million men now in French North Africa, it is difficult to see how the war can be carried on.' He fumed against these 'pusillanimous and defeatist doctrines' and ended by asking 'what Stalin would think of this when he has 185 German divisions on his front'.[2]

Churchill was worried that there might be months during the summer when the Anglo-American armies would be fighting no Germans at all, while Stalin and the Red Army were engaged in a life-and-death struggle with the bulk of the Wehrmacht on the Eastern Front. Furthermore, it seemed to the British Chiefs of Staff that the Americans were directing more landing craft and supplies away from the European theatre and towards the Pacific. Churchill and his entourage, along with the Chiefs of Staff, sailed on the *Queen Mary*, the giant Cunard passenger ship that for several years pre-war had held the record as the fastest across the Atlantic but had now been requisitioned for war work. En route, Churchill and the Chiefs of Staff planned strategy. As they crossed the Atlantic, news came through of the final dramatic victory of the American and British armies in Tunisia. Rommel had got away but nine generals and about 240,000 men had been captured. Churchill ordered the church bells in England to be rung again, for the second time to mark a great victory.

Outwardly, the two weeks of meetings in Washington in May, code-named 'Trident', were a success. But yet again there were serious divisions not far below the surface of Anglo-American relations. Churchill convinced Roosevelt to mediate in the growing dispute between British scientists and their American counterparts, who were now keeping Britain out of the Manhattan Project, the giant scientific research programme to develop an atom bomb. Roosevelt agreed that the deal he had reached with Churchill was to make the results of this research jointly available, a success for the Prime Minister and for British science. The Combined Chiefs of Staff agreed to proceed with the invasion of Sicily now that there had been victory in North Africa, but there was a dispute as to where to go after that. The Americans favoured Sardinia. Churchill and Brooke argued that an invasion of the Italian mainland was the next logical step, with the objective of knocking Italy out of the war. Churchill said this would divert dozens of German divisions not only to fight in Italy but also to replace the Italian divisions currently operating in the Balkans. Here, SOE-encouraged uprisings were taking place and Tito and his Partisans were tying down about twenty enemy divisions. Churchill argued that this was the best support the Allies could give to Stalin during 1943. To his annoyance, the issue remained unresolved. 'I was deeply distressed at this,' he later remembered.[3] However, far and away the most important decision made at Trident was that serious planning should now begin for the invasion of Northern Europe.

As we have seen, the Americans had wanted to launch an invasion of Northern Europe as early as 1942, hoping to prevent a collapse of the Soviet Union and in the belief that victory in Europe could be achieved only by crushing the German armies in ground offensives. They regarded other campaigns as sideshows to this central and necessary assault. Churchill and Brooke, along with the rest of the British Chiefs of Staff, believed that such a complex operation as a cross-Channel invasion should be undertaken only when success was certain. And for success, the Allies first needed to defeat the U-boat menace in the Atlantic and then gain air supremacy over Northern Europe. In addition, the British argued that the army that waded ashore would have to be large enough to sustain the heavy counter-attacks that would be certain to follow. And it would take some time for sufficient supplies of troops and their vital equipment to be built up in southern England. So the British strategy, which had largely prevailed up to the spring of 1943, was to fight more limited campaigns in places where success was more certain, hence the battles in North Africa and the Mediterranean. The British remained sceptical about Operation Overlord, as the invasion plan became known, right up to the end. 'Why are we trying to do this?' Churchill cried out in a depressed moment to Brooke in February 1944.[4] Brooke himself remained fearful up to the eve of the invasion that if the landings failed, it would be the greatest disaster of the war. Nevertheless, at Trident, both Allies committed themselves to the invasion of Europe. A team led by British General Frederick Morgan and known as 'COSSAC' (Chief of Staff to the Supreme Allied Commander) was instructed to prepare outline plans for the invasion of Europe. And there was a date to work to, 1 May 1944, to be known as D-Day.

From Washington, Churchill flew with both the British and the US chiefs, Generals Brooke and Marshall, to Eisenhower's headquarters in Algiers. En route their Boeing Clipper was hit by lightning but not harmed. After the tense atmosphere in Washington, relations improved in the sunshine of Algiers. Marshall was happy to leave open the option of moving on to mainland Italy, depending upon progress in Sicily. Eisenhower impressed Churchill with his sense of authority and command of the situation. And Churchill wanted to ensure that the victorious army now poised in North Africa was kept busy with offensive action over the next year, until the invasion of Europe could be launched. Mostly, the generals agreed with him and he was delighted with the outcome of their meetings. He wrote later: 'I had never received so strong an impression of co-operation and control as during my visit.' After visiting the troops in Tunis and giving a speech to soldiers gathered in the dramatic setting of the ruins of an immense Roman amphitheatre in Carthage, Churchill set off for home. He later recalled: 'I have no more pleasant memories of the war than the eight days in Algiers and Tunis.'[5]

Churchill flew back via Gibraltar, where, because of the weather, he had to leave the luxurious Boeing Clipper and return in the far more basic Liberator. That same day, another plane was flying on a scheduled flight from Lisbon in neutral Portugal back to England. German agents at Lisbon airport noted a large man smoking a cigar boarding the flight and mistook him for Churchill. Luftwaffe fighters were scrambled and shot down the aircraft. All the passengers were killed. Churchill was not on board, but the leading actor Leslie Howard was. His death was a great loss to British cinema. Although Churchill was totally committed to these long-haul flights, each of them had its element of danger.

Back in London after another month away, there were jokes in the press about 'Prime Minister visits Britain'. But Churchill as ever threw himself into a full round of War Cabinet meetings, reports to Parliament and a packed schedule. The invasion of Sicily was launched on 10 July. There were some terrible mishaps at the start when airborne troops were dropped in entirely the wrong places, some of them even in the sea, but overall the landings were a success. Furthermore, many important lessons were learned of value to the later Overlord landings. The US Army advanced from the south of the island to Palermo while the British advanced up the east coast through Catania. General George S. Patton, the US commander, felt intense rivalry with General Montgomery and ended up racing him to Messina. Patton won. In just over five weeks, the whole island was in Allied hands.

Another summit with Roosevelt took place in mid-August in the Canadian city of Quebec, code-named 'Quadrant'. This time, unusually, Clementine accompanied her husband along with their daughter, Mary. At Quebec the same routine followed with the American and British Chiefs of Staff arguing over the details of military strategy while Roosevelt and

Churchill met to endorse the Allies' broad war aims. The Americans were still suspicious of British ambitions in the Mediterranean. And the British continued to press for an attack upon mainland Italy, where Mussolini had now fallen and been replaced by Pietro Badoglio as Chief Minister. But this time the American chiefs were determined not to be out-argued by the British. There was much discussion about policy in the Far East. As we have seen, the British interest here was principally to defend Burma in order to protect India from the Japanese. By contrast, the Americans wanted to use Burma to supply and reinforce the Chinese. In the British delegation was Brigadier Orde Wingate, fresh from leading his 'Chindits' in a long-range, three-month penetration raid behind Japanese lines. This was just the sort of high-risk operation that Churchill loved, and he possibly saw Wingate as a sort of Lawrence of Arabia figure. The Combined Chiefs of Staff agreed to support Wingate in future raids behind Japanese lines. They also agreed to appoint Lord Mountbatten as Supreme Commander in South-East Asia. But Roosevelt and Churchill never really saw eye to eye on the Pacific. Churchill's ambition was to win back all the colonies that had been lost in the Far East. The Americans had no desire to see the war strengthen the British Empire in a region they now regarded as their own.

With regard to Europe, Churchill agreed with Roosevelt that an American general, probably Marshall, should be appointed Supreme Commander for Overlord. Churchill had already offered the post to Brooke but now somewhat abruptly told him that it would be going to an American. He later wrote that Brooke bore his disappointment with 'soldierly dignity'.[6] But Churchill had failed to appreciate what a crushing blow this was to his closest military adviser and colleague. 'He offered no sympathy, no regrets at having had to change his mind, and dealt with the matter as if it were one of minor importance!'

Brooke later wrote. It took him several months to recover from
the loss of a command that he had eagerly sought.[7] This was
not Churchill at his best or his most considerate.

In the discussion about Overlord, Churchill again provoked
American suspicions by suggesting that, if the German strength
in Northern Europe proved too great, the Allies should have 'a
second string to their bow'.[8] He proposed the invasion of northern Norway, an old hobby horse of his. It was not taken very
seriously. More significantly, the Americans proposed an invasion of southern France, Operation Anvil, to coincide with
Overlord. This was approved but would cause much tension
later. In discussions about Overlord, the COSSAC plan was
given preliminary approval. Three divisions of troops were to
land not across the shortest stretch of the English Channel in the
Pas de Calais, where German defences were at their strongest,
but further west, in Normandy, where conditions were better
and defences were weaker.

After Quadrant ended, Churchill took a brief holiday in the
mountains. He quickly relaxed in his lakeside cabin, spending
the days fishing and entertaining his guests at dinner by singing
music-hall songs from his youth along with the latest from
Noël Coward. On 1 September, he returned to Washington
to rejoin Roosevelt, once again living in the White House as the
President's guest. During the course of his stay, Italy formally
surrendered. The first of Hitler's allies had been defeated. At
the end of this visit, Roosevelt had to leave Washington and
placed the White House at the Prime Minister's disposal. Here
Churchill chaired a final meeting of the Combined Chiefs of
Staff. He described this later as a great 'honour' – to preside over
a meeting of the Allied chiefs in the Council Room of the White
House. To him, it was 'an event in Anglo-American history.'[9]

In many ways, this moment was the peak of the special
relationship between Churchill and Roosevelt. But the balance

of that relationship was already changing. Churchill had kept Britain going through a critical period in 1940 and 1941. British soldiers and sailors had borne the brunt of the fighting in North Africa and the Mediterranean until the end of 1942. And British scientists had come up with fabulous technological advances to assist the war effort. But America was rapidly becoming predominant in the war in Europe, with ever more troops arriving in Britain to prepare for Overlord. Within months, there would be one million American GIs in Britain. American bombers were matching the efforts of the RAF in the bombing offensive against Germany. For the Pacific, the US Navy was launching ships on a prodigious scale and American marines and heavy bombers would soon play the dominant role in the war effort there. And as the 'arsenal of democracy', factories in the United States were producing tanks, weapons, planes and other vehicles on a previously unimaginable scale. Meanwhile, the Soviet Union had turned from being an ally facing imminent defeat into a superpower with the strongest army in the world. Like a gigantic steamroller, the Red Army was already driving westwards and engaging the lion's share of Hitler's ground forces.

Moreover, the autumn of 1943 saw the British war effort pushed to its extreme. There were now five million men and women in the armed services. Factories were going flat out. Resources were fully utilised. On 1 November, Churchill circulated a minute that noted: 'Our manpower is now fully mobilized for the war effort. We cannot add to the total; on the contrary it is already dwindling. All we can do is to make within that total such changes as the strategy of the war demands.' Britain was fully stretched and this would inevitably limit Churchill's ambition.[10]

Following the Italian surrender, Allied forces landed on the beaches in the Gulf of Salerno, south of Naples. Then Monty's

8th Army began a slow advance up the east coast of Italy. The Allies had missed the opportunity for a quick follow-up to the capture of Sicily and Hitler had reinforced his army in Italy. German troops moved into Rome, and the whole country was effectively occupied by the Wehrmacht. The six German divisions in Italy in the summer had become twenty-five by the autumn. And with the sudden withdrawal of the Italian Army from the Balkans and Greece, Hitler doubled the size of his garrison there, from twelve to twenty-four divisions. So the collapse in Italy had already achieved the first effect Churchill had wanted – to divert German divisions from the Eastern Front. Stalin sent a message to both Roosevelt and Churchill in September congratulating them and noting that 'the successful landing at Naples and the break between Italy and Germany will deal one more blow upon Hitlerite Germany and will considerably facilitate the actions of the Soviet armies at the Soviet-German front'.[11] On the other hand, the reinforcing of Italy by crack divisions from the East meant that the Allied advance up the 'soft underbelly of Europe', as Churchill called it, soon slowed to a crawl. This revived Marshall's fears that, far from offering up swift conquests, the Italian front would act as a drain on Allied resources and pin down troops that were needed for the bigger mission of Overlord. It had been agreed at Quebec that seven divisions would be withdrawn from Italy to aid in the build-up for Overlord in England in November, and that shipping, particularly landing craft, would be reassigned to the English Channel. But after requests from Eisenhower and an intervention from Churchill, the American chiefs reluctantly agreed to leave most of the landing craft in Italy.

Churchill still hankered after operations in the eastern Mediterranean and the Aegean to buttress Greece and to encourage Turkey to join the Allies. He was keen to seize Rhodes and the

Dodecanese islands along the Turkish coast. 'Improvise and dare,' he instructed General Maitland Wilson, the commander in the region.[12] Churchill argued his case in a long document sent to Roosevelt in October, and ended with a simple appeal: 'I beg you to consider this and not let it be brushed aside.' Roosevelt responded the very next day: 'I do not want to force on Eisenhower diversions which limit the prospects . . . of the Italian operations . . . It is my opinion that no diversion of forces or equipment should prejudice Overlord as planned. The American Chiefs of Staff agree.'[13] This unequivocal response made it quite clear who was in charge now.

Despite the put-down, Churchill was desperately anxious that plans for Overlord should not undermine the campaign in Italy, which he had championed for over a year. He felt that the Americans were being too rigid in their support for Overlord and that he was fighting with his hands tied behind his back. He went back to Roosevelt over and over again during that autumn with his concerns about the forthcoming invasion of France. 'My dear friend,' he concluded one of these messages, 'this is much the greatest thing we have ever attempted. And I am not satisfied that we have yet taken the measures necessary to give it the best chance of success.' Only a few days later in another message he said: 'I am more anxious about the campaign of 1944 than about any other with which I have been involved.'[14] Meanwhile, Stalin began to express his irritation with the slow progress of the Allied armies in Italy. He protested that, having fortified Italy, Hitler was now sending units back to the Eastern Front.

With these dilemmas and arguments in mind, at the end of November the Allied leaders, this time including Stalin, agreed to meet in Teheran for the first full summit of the war, the first of the 'Big Three' meetings. Churchill sailed from Plymouth on HMS *Renown* on 12 November. He would be away for two months and travelled with his usual team. First, there was the staff of his Map Room. Here, Captain Richard Pim constantly updated wall charts with pins and lines so Churchill could keep abreast of daily movements on every front and in every ocean. Pim packed up the Map Room and set it up wherever the Prime Minister was based. A complete signals unit travelled with him too, so Churchill could be kept in daily or even hourly contact with London, and through which he could send daily summaries to Deputy Prime Minister Attlee and other ministers. The vitally important Ultra reports were also sent to him by secret communication links to keep him updated. And, of course, there was the small group of secretaries who accompanied Churchill everywhere with their silent typewriters, ready and waiting at any hour of the day or night to type up minutes, briefing papers, speeches or any other message that the Prime Minister wanted to dictate. It is a sign of the success and stability of Churchill's regime that he could afford to travel and operate abroad for such long periods of time. Neither Hitler nor Stalin could have imagined being so far from his power base for so long.

The meeting in Teheran was preceded by another conference between Roosevelt and Churchill and their military chiefs in Cairo, code-named 'Sextant'. They wanted to get their position clear before presenting it to Stalin and his chiefs for the first time. But this meeting was unlike their previous conferences. Roosevelt set the agenda, literally. Much time was spent discussing the Far East and the Pacific. The presence of the Chinese Nationalist leader, Chiang Kai-shek, also meant the conference focused much more on the East. Plans were made for a big amphibious landing in the Bay of Bengal, to help provide supplies for China. This was to be called 'Operation Buccaneer'. Tensions between the British and American chiefs, never far below the surface, erupted in some violent exchanges.

In the afternoon session on 23 November, Admiral Ernest King and General Brooke argued particularly vehemently about the supply of landing craft. The American General Joseph Stilwell, who was present, wrote: 'Brooke got nasty and King got good and sore. King almost climbed over the table at Brooke. God he was mad! I wish he had socked him.'[15] Many unresolved and uncertain issues were taken on to Teheran. The prospects for a successful summit did not look good.

By the time the Big Three gathered in Teheran on 28 November, Churchill had gone down with a bug. He had left London two weeks earlier with a heavy cold and a sore throat. Now, after many late-night sessions in Cairo, he had, perhaps symbolically, lost his voice. His doctor, Charles Wilson, offered Churchill sprays to restore his vital weapon. Then, at the first dinner hosted by Roosevelt, the President had to withdraw with stomach ache. There were even rumours that he had been poisoned. But despite this inauspicious start, the summit proceeded reasonably well and after some tough talking, several significant decisions were reached. At the first plenary session, having been briefed about Overlord, Stalin surprised both Roosevelt and Churchill by eagerly picking up on the plans for Anvil, the invasion of southern France. No doubt the Soviet leader was already thinking of the post-war world, in which he had no desire to see Anglo-American forces in the Balkans. He wanted that part of South-East Europe in his own sphere of influence. By supporting Anvil, he probably believed he would divert Allied forces away from the Balkans. There was a discussion over the timing of Overlord. The Russians insisted on May 1944. The Americans and the British were now planning on June. Churchill still argued his standard line that he did not want to undermine the campaign in Italy, where there were twenty British or British-controlled divisions, simply to guarantee a 1 May date for Overlord.

Stalin kept up the pressure on the President and the Prime Minister. At the next plenary session, he asked who was going to command Overlord. Roosevelt responded that this had not yet been decided. Stalin retorted that the operation 'would come to nought' unless one man was placed in charge of both preparing and leading the invasion. And when Churchill spoke about the need to ensure that German forces in France were not strong enough to throw the Allies back into the sea, Stalin asked bluntly if 'the Prime Minister and the British staffs really believed in Overlord?' Churchill responded that, as long as conditions were right, British forces would hurl everything they had across the Channel at the Germans with 'every sinew of our strength'.[16] But Stalin kept pressing that the invasion must be launched in May. No delay was acceptable.

Roosevelt wanted to build up his own rapport with Stalin in Teheran, which inevitably meant distancing himself from Churchill. As the President put it, he could not appear to be 'ganging up with the British'. Churchill understood what was going on but sought to reassure Stalin during a personal meeting that he was entirely committed to Overlord, that he had no wish to intrude into the Balkans, and that his plans in Italy were limited.

Then, at the second dinner, hosted by Stalin, an unusual incident occurred. The discussion turned to the subject of punishing the Germans at the end of the war. Stalin suggested that there were fifty thousand leading Nazis who were really behind the whole German war effort, and when the war was won they should all be shot. Churchill was genuinely shocked and responded: 'The British Parliament and public will never tolerate mass executions.' Stalin persisted and Churchill said he would rather be taken out into the garden and shot there and then than sully his country's honour with such infamy. Roosevelt, trying to diffuse the brewing row with humour, suggested a compromise – only 49,000 need be shot. Then, to everyone's surprise, the President's son Elliott, who was attending the summit as a military attaché, rose to his feet. No doubt under the influence of the fine wine, he said the US Army would support Marshal Stalin's plan. Churchill, who was so used to being in command of the table talk, got up in disgust and walked out of the banquet and into an adjoining room. Only a minute had passed before he felt a hand on his shoulder. It was Stalin, grinning away, saying it had all been a joke and he had meant none of it. 'Stalin has a very captivating manner when he chooses to use it,' Churchill wrote later. The Prime Minister returned to the table and the rest of the dinner passed off pleasantly enough.[17]

Churchill was feeling the strain of losing the President's friendship and support and of being in the unfamiliar position of not getting his own way. Later that evening, Charles Wilson found him in a very depressed state, brooding on an apocalyptic vision of post-war Europe. The doctor took Churchill's pulse and told him it was high, probably because of the drink. 'It will soon fall,' mused Churchill. Then, returning to his gloom and despondency, he asked: 'Why do I plague my mind with these things? I never used to worry about anything. Stupendous issues are unfolding before our eyes, and we are only specks of dust, that have settled in the night on the map of the world.' He then turned to his doctor and asked abruptly: 'Do you think my strength will last out the war? I fancy sometimes that I am nearly spent.' Then he got into bed. The doctor waited a few minutes and asked if he wanted him to turn out the light. There was no answer. Churchill was already asleep.[18]

The following day, Stalin seemed more positive. He agreed to launch a major offensive on the Eastern Front to coincide with Overlord, in order to prevent German troops being diverted to the Western Front. He also announced that once

Germany had been defeated, the Soviet Union would join the other two Allies in their assault upon Japan. This transformed planning for the final stages of the war in the Far East. Roosevelt confirmed that Overlord would be launched in May 1944. The three leaders also discussed the value of deception tactics to confuse the enemy. Churchill agreed that 'truth should always be attended by a bodyguard of lies'.[19] And so Operation Fortitude was born, the secret deception plan for Overlord.

On this third evening, 30 November, it was Churchill's turn to host dinner. It was his sixty-ninth birthday. General Ismay remembered that the toasts and short speeches began as soon as everyone had sat down. In the Russian style, they then continued throughout the evening. Churchill spoke about the President's fine achievements and called the Russian leader 'Stalin the Great'. As the evening progressed, the toasts got merrier. At one point Churchill proposed a toast to the 'proletarian masses' and Stalin responded with a toast to the 'Conservative Party'! For Churchill, it was a memorable birthday: 'On my right sat the President of the United States, on my left the master of Russia. Together we controlled a large preponderance of the naval and three quarters of the air forces in the world, and could direct armies of nearly twenty millions of men, engaged in the most terrible of wars that had yet occurred in human history.' Later, he would describe the evening more amusingly: 'There I sat with the great Russian bear on one side of me, with paws outstretched, and on the other side the great American buffalo, and between the two sat the poor little English donkey who was the only one . . . who knew the right way home.'[20]

Back in Cairo after Teheran, Roosevelt took a momentous decision. He decided to appoint Eisenhower rather than Marshall as Supreme Commander for Overlord. The President explained this to Marshall by claiming that he would not be able to sleep at night if the general were to relinquish his role as Chief of Staff and be out of the country. If he were disappointed, Marshall did not show it. He said he would go along with whatever his commander-in-chief wanted. Roosevelt told Churchill of his decision during a short sightseeing visit to the Sphinx, a friendly and relaxed interlude after many days of tension. The two leaders gazed at the ancient statue for some minutes as the evening shadows fell, as if seeking guidance. But, as Churchill later put it, the Sphinx 'told us nothing and maintained her inscrutable smile'.[21]

Churchill was exhausted by the tensions of the previous weeks. General Smuts, the South African leader, told Brooke that he thought the Prime Minister was working too hard and that he was 'beginning to doubt whether he would stay the course'. Churchill himself admitted that he felt very tired and noted 'that I no longer dried myself after my bath, but lay on the bed wrapped in my towel till I dried naturally'.[22] He flew on to Eisenhower's headquarters at Carthage, outside Tunis, where he collapsed in exhaustion and went to bed.

At about four o'clock the following morning, 13 December, General Brooke was fast asleep when he was awoken by someone in his room mournfully calling out, 'Hulloo, Hulloo, Hulloo!' The most senior officer in the British Army leapt out of bed and turned on his torch to find the Prime Minister wandering around in his dressing gown with a brown bandage wrapped around his head. He was confused and thought this was his doctor's bedroom. When he finally found Charles Wilson, he was once again diagnosed with pneumonia. He was given a brand-new antibiotic sulphonamide called M&B that had been launched only recently on the market. But over the next few days Churchill's health rapidly deteriorated. Specialists were flown in from Cairo. On the night of the 15th, Churchill's heart began to fibrillate and his pulse became erratic. His doctor was seriously worried and thought this time he might not pull through. Churchill told his daughter Sarah, 'If I die, don't worry – the war is won.' John Martin, the Cabinet secretary who was accompanying him on this trip, alerted the War Cabinet in London. Grave announcements were issued to the press. At Carthage, Sarah read extracts from Jane Austen's *Pride and Prejudice* to her father. Clementine flew out from London to be with her husband. Then, suddenly, on the 17th, the antibiotics finally started to do their bit. From then on, Churchill recovered a little more each day. He started to dictate messages from his bed again and saw a constant stream of visitors until Clementine said all she could do was 'poke my nose around the corner of the door'. Churchill soon became irascible with his doctors, but they insisted he have at least a week's rest. So he spent a few more days in Carthage.

On Christmas Day, his first day out of bed, he had a conference with Eisenhower, Wilson, Alexander, Tedder and Cunningham to discuss plans for the landings at Anzio, south of Rome. Again the shortage of landing craft threatened the future of the whole operation. Churchill argued that the landings were essential to the Italian campaign, and the others agreed that they must go ahead, even if this led to a delay for Overlord. Two days later, Churchill and his entourage travelled on to the same villa in Marrakech where he and Roosevelt had watched the light over the mountains eleven months before. Churchill rested there for three weeks. He had suffered a serious, life-threatening illness and it had been an anxious time for everyone around him. But the old bulldog had pulled through.[23]

Throughout his illness, Churchill had been in contact with Roosevelt to agree the senior commanders for Overlord. Eisenhower was to take up his post as Supreme Commander as soon as he could leave his Italian command. Tedder was to be his

deputy. And Montgomery was appointed Land Commander.
Monty rushed back to London, leaving the 8th Army in Italy.
On 2 January 1944, he attended a conference at St Paul's School
(his old school in west London), which had been requisitioned
by COSSAC, the group that had been quietly making plans for
D-Day since the previous summer. After a briefing on the outline of the plans so far, Monty stood up and spoke. He tore the
plans to shreds and said that the invasion was on too narrow
a front with forces that were too small. He sent the COSSAC
staff back to the drawing board to draw up a new and more
ambitious plan. Later, Monty would claim that the redrawing
of the invasion plans for Overlord was all his own doing. In
fact, he and Eisenhower had already discussed the issue and
agreed their line. And they knew that Churchill also thought
more divisions were needed. The truth was that COSSAC had
gone so far, but now that the Supreme Commander and his
field commanders were in place, the whole operation could be
scaled up. Serious planning for the invasion began at this
point.

Eisenhower arrived in London in mid-January, and on the
21st he presided over the first meeting of his commanders
and planners from the office that became known as Supreme
Headquarters, Allied Expeditionary Force (SHAEF). From here
the battle plan for the Overlord invasion soon expanded, as
Eisenhower had the authority to convince Washington of the
need for extra landing craft, air support, naval escorts, and the
troops and *matériel* vital to launch a successful invasion across
a hundred miles of the English Channel.

The invasion of Northern Europe was one of the most ambitious and risky operations of the Second World War. The scale
of the challenge facing the planners was genuinely awesome.
Each armoured division required forty ships to transport it. The
Americans assembled in the South-West and the British troops

concentrated in the South-East of England. Every unit required training camps, assembly areas and acres of space to gather their vehicles, armour and weaponry. A total of 137,000 wheeled and semi-tracked vehicles, 4000 full-tracked vehicles and 3500 artillery pieces were brought across the Atlantic in supply ships for the American armies alone. Jeeps from Detroit, K-rations from the farms of the Mid-West, shells, weapons, ammunition and medical supplies from across America were all delivered to the units assembling in England. Stretches of the Devon coast were evacuated of all their inhabitants so practice landings could take place using live ammunition. Ships and landing craft were lined up in all the major ports along the south coast of England. Hundreds of thousands of aerial photographs were taken up and down the French and Belgian coast and distributed so the troops could study the objectives that would face them. Divers crept ashore at night to collect samples of sand to calculate if it could support armoured vehicles, to measure the slope of the beaches and to assess the shore defences.

The final battle plan for D-Day was for three divisions of airborne troops to be dropped behind enemy lines to seize key targets on both flanks in the night before the invasion. During the morning of D-Day itself, four army corps would land on five Normandy beaches from the Cotentin peninsula in the west, on the right flank of the invasion, to the Orne River on the east. US, British and Canadian troops would charge ashore at dawn, and during the day about 170,000 men, 20,000 vehicles and thousands of tons of supplies would arrive to support them. Operational plans were drawn up and hundreds of pages of orders were circulated at corps level, then for divisions, brigades, regiments and companies. The five beaches were given code names, from west to east: Utah, Omaha, Gold, Juno and Sword. Each beach was divided into subsections and each unit assigned its targets from H-Hour, the moment when the first wave hit the beaches, onwards.

Churchill was too busy to engage with the day-to-day planning of the largest and most complex military operation of the war. But he still wrote minutes and issued his thoughts on matters like the waterproofing of vehicles, the onshore bombardment from naval ships and the number of transport aircraft available for the airborne assault. His influence was really felt when it came to his support for the Eisenhower–Montgomery plan to broaden the bridgehead and increase the number of men who would land ashore on D-Day. This would mean delaying the date of the invasion and reneging on the promise Roosevelt had made to Stalin at Teheran. The argument dragged on for a couple of months, until late March, when Washington finally agreed to the revised plan. Once that decision was made, and the idea of a May attack was abandoned, the next point when the moon and the tides would be right for a beach landing was from 5 to 7 June.

Churchill attended two presentations by Monty at St Paul's School, outlining the developing plans for Overlord. At the first on Good Friday, 7 April, Churchill was not well and he seems to have contributed little. The second, on 15 May, was a big show with dozens of senior officers from the British and American armies sitting in rows on benches, and the King, Churchill, Smuts, Brooke and the other Chiefs of Staff in VIP chairs at the front. Churchill was impressed by what he heard and said he was 'hardening' to the Overlord project.[24] Monty spent the last few weeks before the invasion on a whirlwind tour of southern England, inspecting the troops and visiting factories. His clipped, matter-of-fact style, captured in some of the newsreels, comes across as rather comical today. But it proved genuinely popular at the time. He addressed people in their own language, made them laugh and made them feel that they were really contributing to something important. And he seemed like a real soldier, rather than some aloof and remote château-general. The government was wary about Monty emerging as national figure, a popular warlord, but Churchill never did anything to prevent him from drumming up support.

As we saw in the last chapter, Churchill lost the struggle to keep RAF Bomber Command out of the planning for D-Day. Eisenhower brought all the British and American bombers in Britain under the command of SHAEF to enable strikes at tactical objectives along the beaches and at transport hubs inland. In order not to reveal that Normandy was the objective, this bombing had to be carried out along the whole French and Belgian coast.

Churchill's influence, however, was still considerable in important aspects of the Normandy invasion. A serious blot on his reputation for many years had been his involvement with the Dardanelles fiasco in 1915. Thousands of men had been landed along the coast at Gallipoli and it had been a near disaster. Obviously, Churchill did not want to be associated with the repetition of anything like this. And memories of the horrible slaughters resulting from First World War human-wave assaults against well-defended positions were strong in the minds of Churchill and the British generals, even if they were not for the American planners. As we know, from his early days, Churchill had been fascinated by new weapons and armoured vehicles of war, such as the tank which he had contributed to the development of in 1915. And in May 1940 he had asked Cherwell to review the production of tanks and called for the construction of an additional one thousand of them. One of the tank men who had given Churchill useful information during his 'wilderness years' about the parlous state of British armoured regiments was General Percy Hobart, then one of the most experienced commanders of armoured units in the army.

Churchill had been impressed with his fertile mind and his energy. But Hobart was a prickly figure and unpopular with his seniors in the army. In the summer of 1940, Churchill was appalled to discover that the Army High Command had pensioned off Hobart and he was serving his country only as a lance-corporal in the Home Guard. Churchill immediately called for his reinstatement as a major-general, and in a minute to his CIGS he wrote of Hobart's 'strong personality and original view'. He concluded in a way that perfectly summed up his view on these matters: 'We are now at war, fighting for our lives, and we cannot afford to confine Army appointments to persons who have excited no hostile comment in their career . . . This is a time to try men of force and vision and not to be exclusively confined to those who are judged thoroughly safe by conventional standards.' More prosaically, he was overheard saying to Dill: 'Remember it isn't only the good boys who help to win wars; it's the sneaks and stinkers as well.'[25]

Hobart was brought back into the army and given command of what became the 11th Armoured Division. He reorganised the structure of the unit to include a broad mix of heavy and light armour along with artillery, and can lay claim to having been the architect of the modern armoured division.[26] Churchill organised a series of four 'Tank Parliaments' at Downing Street in May and June 1941, another element in the War Lab. At these meetings, leading military figures and civilian experts met to exchange views about the future use of tanks, and of course to listen to Churchill's own thoughts on the subject. Hobart made an important contribution at these sessions. It is no surprise that the latest thirty-ton tank to roll off the production lines in 1941 was named the 'Churchill'. But the tank's namesake was distressed to discover that the mark-one Churchill had only a two-pound gun that was woefully inadequate for armoured warfare. It was unusual for anything with the name 'Churchill' to be known for its lack of bark. The gun was subsequently upgraded to a six-pounder.

In April 1943, Hobart was put in charge of the newly formed 79th Armoured Division, which became known as the 'Zoo' or 'Menagerie'. His task now was to devise a range of armoured devices to assist landing troops in getting through beach obstacles and in capturing the beachhead. The whole stretch of coast from Brittany to Norway had been heavily fortified. Concrete pill-boxes had been built with powerful heavy machine guns to provide arcs of fire across the beaches below. A huge array of steel obstacles had been set up along the beaches, and minefields had been laid in the sand dunes and at the head of each beach. In total, four million mines had been laid. The whole defensive line was called the 'Atlantic Wall'. In January 1944, in a strange twist of fate, Hitler appointed Field Marshal Rommel to take charge of it. Once again, Monty would be up against his old adversary from the desert war. With great vigour and energy, Rommel toured up and down the coast, reinforcing and building up the defences. It was Hobart's job to find the armoured machines that could penetrate the Atlantic Wall and support the infantry wading ashore.

Hobart turned his new division into a sort of think-tank for armoured warfare. Everyone was encouraged to come up with ideas. Floating tanks, known as 'Duplex Drives' (or DDs), had already been developed in the Mediterranean. They could be launched at sea with floating skirts around them so they could swim in with the landing craft carrying the infantry. Hundreds were built for D-Day. Hobart and his men devised minesweeping flail tanks with a giant rotor that spun a set of whirling chains on the ground to detonate mines ahead of the advancing tank. There were Bobbin tanks, which carried a metal track above the turret. Through an ingenious feeder mechanism, this could be laid in front of the advancing tank to cover ground where the sand was too soft to support tank tracks. Then there were tanks armed with flame throwers to flush out the enemy from well-defended bunkers. Tanks were adapted to lay bridges. And there were bulldozer tanks to remove obstacles. Hobart's inventiveness turned the British into pioneers of specialised armour, and the machines he devised became known as 'Hobart's Funnies'. Churchill was delighted with them. The Americans were more sceptical and used only the floating DD tanks on D-Day. Brooke wrote in his diary after a visit to Hobart to inspect the assault vehicles: 'Hobart has been doing wonders in his present job and I am delighted that we put him into it.'[27] Churchill's faith in Hobart had fully paid off.

Another aspect of the challenge of landing an army on beaches had exercised Churchill's mind for some time. A practice assault on an enemy-held harbour had been tried at Dieppe, in August 1942. It proved a disaster, with massive loss of life among the largely Canadian assault force. The lesson learned was that it was going to be very difficult to capture a heavily fortified port from the enemy. With thousands of tons of supplies needing to be brought ashore every day, the solution seemed to be for the Allies to build their own harbours along the invasion beaches. As early as May 1942, Churchill had sent a historic memo to Lord Mountbatten, who was then Chief of Combined Operations, about the problems of building a harbour on a shallow-water invasion beach. This included a handwritten comment that said the piers 'must float up and down with the tide. The anchor problems must be mastered. Let me have the best solutions worked out. Don't argue the matter. The difficulties will argue for themselves.'[28] This was typical of Churchill, and the last three sentences are still quoted in management training schools today as the attitude to adopt when facing a substantial challenge.

The War Office took on responsibility for pursuing this further, and Brigadier Bruce White was put in charge of the project. When Churchill sailed to the Quebec Conference on the *Queen Mary*, he debated the issue with his senior commanders. It had been decided to use blockships and giant concrete caissons to protect the harbours from the full force of the sea. Churchill did not understand how this would work, so an impromptu demonstration was staged in his bathroom. Several paper boats were launched and the water was splashed about. The paper boats sank. When a barrier was placed across the bath to represent a breakwater, no amount of splashing sank the little boats at the other end. General Ismay recalled:

> If a stranger had visited his bathroom, he might have seen a stocky figure in a dressing gown of many colours, sitting on a stool and surrounded by a number of what our American friends call 'Top Brass', while an admiral flapped his hands at one end of the bath . . . and a brigadier stretched a lilo across the middle . . . The stranger would have found it hard to believe that this was the British High Command studying the most stupendous and spectacular amphibious operation in the history of war.[29]

Churchill was now convinced it would work. The harbours were code-named 'Mulberry'. Two of the giant structures were to be built in Britain, one for the Americans off Omaha Beach and one for the British at Arromanches, off Gold Beach.

In a normal harbour, the quay remains fixed and the boats float up and down with the tide. In the Mulberry harbours, the piers had to float up and down with the ships. Giant pontoons were designed to be sunk on the seabed with four steel legs one hundred feet high. The floating pierheads would then be linked to the shore by a roadway strong enough to support thirty-ton tanks and other heavy vehicles. Major Alan Beckett, a distinguished engineer, came up with the design that combined both strength and flexibility using steel cables and heavy-duty ball-and-socket joints. A new form of anchor was also designed to secure all this to the seabed. And to protect it, huge concrete blockhouses, the caissons, each weighing six thousand tons, would be built in Britain and then floated across the Channel and sunk alongside a set of obsolete ships to form a harbour wall against the sea. The scale of the Mulberry project was gargantuan: 45,000 workers produced 147 caissons, 23 pierheads and 10 miles of floating roadway. This placed a near-impossible strain on British industry, which was already stretched to its limit.

In addition to the industrial challenge faced in the construction of the Mulberry harbours, there was endless bickering between the army, who had led the way on the project, and the navy, who regarded it as a matter for them. Churchill's constant requests for updates on the progress of his pet project no doubt prevented it from grinding to a complete halt. Despite endless delays, the many different parts were completed on time in ports and shipyards all around Britain. Like a giant jigsaw puzzle, the separate elements were then towed across the Channel and into position, and the two harbours were assembled a week after D-Day. It was an astonishing achievement on every level, and Churchill's enthusiasm and persistence had been vital to keep it going.

Throughout the whole planning for D-Day, the major fear was of a breach of secrecy – that somehow the plans would get out and the Germans would be waiting. Part of the attempt to prevent this was that huge operation of deception that had been born in the conversations with Stalin at Teheran. Through Operation Fortitude, the Allies tried to persuade the German

High Command that the real invasion would be launched along the shortest stretch of the Channel from England, on the Pas de Calais, and that the Normandy landings, when they came, were only a diversion. To this end, a false army known as '1st Army Group' was created in south-eastern England, complete with dummy tanks, trucks and fleets of dummy ships in the ports of Kent. A complete signals unit sent endless messages about training and assembly that could be picked up by the Germans. The 1st Army even had its own commander, the bullish General Patton, whom it was thought the Germans would believe was the most likely person to command the invasion forces. Patton made himself as visible as possible to add to the deception. Again, Churchill loved this sort of operation and gave it his full support.

Meanwhile, the advance in Italy, of which so much had been hoped by Churchill and others, had slowed to a snail's pace. The Germans had heavily fortified a defensive position known as the 'Gustav Line' across the central mountains, focused upon Monte Cassino, which dominated the road north to Rome. For four months, the Allies tried everything to destroy this citadel, including bombing the medieval Benedictine monastery that stood on its peak, destroying a library and antiquities that were over a thousand years old. But it was not until 18 May that Polish troops were able to capture the mountain that dominated the surrounding countryside. Finally, the troops advancing north met up with those who had landed at Anzio. They pressed onwards together, and on 4 June General Mark Clark led the Americans into Rome. It was splendid news which cheered Churchill and Roosevelt.

With the date for the invasion of France now fixed for 5 June, everything slowly came together. The men were trained and ready. They were in their final assembly camps near the embarkation points and received their final briefings. They were then issued with ammunition, seasickness pills and a leaflet about how to behave towards French civilians. From this point, no one was allowed to leave their base. Four thousand landing craft and hundreds of assault vehicles, along with dozens of escort ships, were ready. Thousands of light and heavy bombers were waiting to blitz the beach defences. The process of embarkation began. Anti-aircraft defences and fighter planes were on full alert, ready to attack any Luftwaffe bomber that dared to intervene. From Ipswich, along the whole south coast of Britain, and right round to Bristol, southern England had become an armed camp waiting for the order to go.

As we have seen, Churchill was eager to watch Overlord from HMS *Belfast*. But his senior commanders were horrified by the thought of the Prime Minister's being present near the landing beaches – not only for the danger it posed to him, but because of the fear that he might try to intervene at some critical juncture. In the end, to their relief, the King instructed Churchill to remain in England.

Then, at the last minute, the weather intervened. A storm blew up in the Atlantic and Eisenhower was advised that this might cause chaos for shipping and air support. At a late-night meeting on 3–4 June, he agreed to postpone D-Day for twenty-four hours. With 200,000 men in this state of readiness, it could be fatal to delay much longer. In the early hours of 5 June, Eisenhower gathered again with his commanders at his headquarters near Portsmouth. The predicted storm was raging and rain was lashing in horizontal streaks. The chief meteorologist, Group Captain Snagg, reported that he had spotted a slight break in the bad weather out in the Atlantic. He thought it would last for about thirty-six hours. Eisenhower had to make the biggest decision of his life. He turned to Montgomery and asked his opinion. He asked his deputy, Tedder, and the other senior commanders. The Supreme Commander listened carefully to their answers but knew that that only he had the authority to take the final decision. Eisenhower paused, then said, 'Let's go.' It was 4.15 a.m. on 5 June. The invasion was on.

Churchill, who had been sceptical of Overlord for so many months, had by now been won over and was optimistic. He invited the Chiefs of Staff to lunch and told them that the invasion was likely to be a success. Brooke, on the other hand, scribbled in his diary on the night before the invasion: 'I am very uneasy about the whole operation. At the best it will fall to very very far short of the expectation of the bulk of the people, namely all those who know nothing of its difficulties. At the worst it may well be the most ghastly disaster of the whole war. I wish to God it were safely over.'[30] Eisenhower chose that evening to visit the men of the 101st Airborne Division at Greenham Common airbase as they prepared to emplane for their drop behind enemy lines. His advisers had predicted they might suffer up to 80 per cent casualties, but he was cheered by their courage and enthusiasm. 'Now quit worrying, General,' one of them said, 'we'll take care of this thing for you.' Eisenhower had in his top tunic pocket a short, handwritten letter of resignation, accepting full responsibility for the failure of the invasion should everything go wrong.

At about 10 p.m. on the night of 5 June 1944, the aircraft carrying the airborne troops began to take off on their missions. They flew in a tight formation in a huge air armada across the Channel. In the west, the transport pilots encountered heavy anti-aircraft fire as they crossed the Cotentin Peninsula. In panic, many of the pilots pressed the green light, the signal to jump, and American paratroopers leapt out of the transport aircraft many miles from their intended drop zones. Some landed so far off course they could not even find where they were from their maps. In the east, some of the British glider troops, by contrast, landed within yards of their target, Pegasus Bridge. In minutes, they had seized control of Pegasus and another key bridge over the Orne. The first blood had been drawn and the 'Longest Day' had started.

The amphibious landings took place soon after dawn on 6 June. At the eastern end of Sword Beach the landing drill worked well. The DD tanks beached successfully and the flail tanks, known as 'Crabs', cleared the minefields as the men moved up and off the beach. The biggest problem came from traffic jams and bottlenecks as so many men and vehicles became entangled trying to get ashore. On Juno Beach, the Canadian landing went more slowly, but by late morning the bulldozer tanks were ashore and clearing obstacles, and the infantry advanced off the beach. After a few hours of fighting, the first coastal villages were liberated. At Gold Beach there was much heavier resistance and many of the tanks were taken out by the German anti-tank guns. But again the Crabs flailed through the minefields and the Bobbin tanks laid paths for vehicles and infantry to move forwards and capture the German strongholds. In total, the British and Canadians lost only 32 assault tanks out of 170. Hobart's 79th Armoured Division lost a total of 179 men killed or wounded. These numbers were way below the predicted casualty rates.

To the west on Omaha Beach, the story was entirely different. This was always going to be a tough landing, with two-hundred-foot cliffs at the head of the beach and well-prepared defenders inside thick concrete bunkers. A new German unit, the 352nd Division, had been assigned to this stretch of coast just a few days before the invasion. They were tougher and more determined than many of the other units in Normandy. At Omaha, the tanks were due to get to the beach at H-5, five minutes before the infantry arrived in their landing craft. But the sea was much rougher here in the aftermath of the storm. Many of the floating tanks were launched too far out and sank straight to the bottom. In one group, 29 out of 32 Sherman tanks sank on launching. And when the infantry waded ashore they were met with a furious enfilade of machine-gun and artillery fire. From their well-dug-in nests, the German machine gunners sprayed arcs of withering fire across the beach. Many men in the first wave never even got off the ramps of their landing craft. Others fell into the sea and drowned under the weight of all they were carrying. As landing craft were hit by mines or shells they blocked the beach for the next wave trying to get through behind them. The best the survivors could do was to take shelter behind the iron beach obstacles or under the sea wall. But moving off the beach under the intense fire seemed impossible. Casualties mounted to alarming levels. The first wave was almost entirely wiped out. Watching the massacre unfold from USS *Augusta* several miles out to sea, General Omar Bradley and his commanders tried to piece together what was going on and considered abandoning the landing. Naval destroyers came in as close to the beach as they dared, within about eight hundred yards, to fire directly into the German gun emplacements. Then, in the late morning, Brigadier Norman Cota, the energetic deputy commander of the 29th Division, showing great heroism, led a group of men up a gully and on to the cliffs above the beach. From here, they were slowly able to fan out and attack the German gun emplacements one by one. By early afternoon, the determination of the US infantry had won through. They had overpowered the German strongholds and cleared the beach exits. But by then there had been three thousand casualties on Omaha Beach.

On the westernmost beach, Utah, by contrast, things went extremely well. The current carried the landing craft more than a mile from their planned landing zone, but the beaches the American 4th Division hit were lightly defended and most of the armour got ashore without incident. Astonishingly, the casualties here during D-Day were lighter than on their last training exercise at Slapton Sands in Devon.

Overall, D-Day was a remarkable triumph. By the end of the day, 177,000 men and their equipment were ashore. Some units had penetrated five miles inland. The Allied air forces had flown 14,600 sorties. The Luftwaffe had barely put in an appearance all day. U-boats had got nowhere near the vast naval armada in the Channel. Hitler's supposedly impregnable Atlantic Wall had been breached. Not anticipating that the Allies would invade during the storm, Rommel had returned to Germany and so was not present when the landings took place. Other officers were away at an exercise in Rennes. Rommel rushed back later in the day, but only Hitler could order the deployment of the key panzer reserves. He believed that Normandy was just a side show, with the real landings to follow in the Pas de Calais. Operation Fortitude had been a success. So, in the first few critical hours, the panzer reserves waited but did not intervene.

Churchill followed events during the morning of 6 June in his Map Room. Later that day, he addressed a packed House of Commons. He paid fulsome tribute to the 'ingenious modifications of the British armour' and said:

> This vast operation is undoubtedly the most complicated and difficult that has ever taken place ... Nothing that equipment, science or forethought could do has been neglected and the whole process of opening up this great new front will be pursued with the utmost resolution both by the commanders and by the United States and British Governments whom they serve.[31]

He sent a message to Stalin, informing him of the initial success of the invasion. Stalin replied with wholehearted praise for the 'grandiose scale' on which the invasion had been carried out. As per the agreement at Teheran, on 10 June the Red Army's vast summer offensive began with an assault on the Leningrad front.

Churchill was still itching to get across to Normandy to see the bridgehead for himself. He asked Monty if he could pay a visit, saying: 'We do not wish in any way to be a burden to you or on your headquarters . . . We shall bring some sandwiches with us.'[32] On 12 June, Churchill crossed the Channel in a destroyer. Brooke, who accompanied him, noted: 'We continually passed convoys of landing craft, minesweepers, bits of floating breakwater being towed out, parts of the floating piers etc. And overhead, a continuous flow of planes going to and coming from France.' They were taken ashore in a DUKW, a floating truck, and were photographed and filmed as they disembarked on to the beach. Brooke was quite emotional at being back in France four years after he had left in the disastrous defeats of 1940. Monty met them and took the party to his headquarters about five miles inland, where he explained his dispositions and his plans. Then they had lunch in the grounds in a tent. Churchill asked how far away the enemy was. Monty told him about three miles and explained that there was not a continuous perimeter line. Churchill asked: 'What is there then to prevent an incursion of German armour breaking up our luncheon?' Monty replied that he didn't think they would come. Churchill and Brooke finished their tour by sailing up and down the invasion beaches and watching an LCT (Landing Craft Tank) disgorge its cargo of tanks and trucks on to the shore 'in a remarkably short time'. Finally, they witnessed a bombardment by two British battleships of positions about twelve miles inland. Churchill had never been on a Royal Navy ship firing in anger, so he asked the captain of his destroyer to fire off a salvo as well. This he did, and Churchill was quite disappointed when the enemy did not return their fire.[33]

Churchill and his team returned to Portsmouth that evening, having seen some of the giant pieces of the Mulberry harbours being assembled. When it was finally built, protected by the giant concrete caissons and sunken ships, with its floating pierheads and miles of roadway, the British harbour became known, fittingly, as 'Port Winston'. Although severely damaged in a storm, it continued to process literally millions of men, tens of thousands of tanks and other vehicles, and hundreds of thousands of tons of supplies for five months, until the Allies had advanced into Holland. The remnants of the giant concrete blocks can still be seen in the sea off the beach at Arromanches, the remains of one of the most extraordinary military engineering feats of the war.

The day after Churchill returned to London from his sightseeing trip to the invasion beaches, a mysterious new type of bomb exploded in a street in Bethnal Green, east London. Six people were killed and another nine injured. Hitler had been threatening to use secret terror weapons against Britain for some time. German science had advanced way ahead of what was happening in Britain. A campaign now began in which hundreds of flying bombs and then ballistic missiles were fired against London. Hitler was not defeated yet.

10

Victory and Defeat

On 15 May 1942, an aerial reconnaissance Spitfire was flying fast and high across the Baltic. Over the northern end of the forested island of Usedom, the pilot noticed a mass of new construction around what looked like an airfield. He turned his cameras on and took a series of photographs of the ground below. When they got back to the aerial photography interpretation centre located in a big country house at Medmenham in Buckinghamshire, the photo interpreters tried to work out what was going on. Medmenham was to aerial photography what Bletchley Park was to code-breaking. An unlikely group of academics, scientists, archaeologists and air force types had been assembled and were building up a detailed analysis of everything that was happening in occupied Europe. By constantly comparing new photos with previous ones, they were able to plot the building or extension of every new factory, every new stretch of road or railway, and every new gun emplacement across Europe. But the strange shapes in the photos taken along the Baltic coast, which included circular embankments, left the interpreters puzzled. It was decided to keep a close watch on developments in this place called Peenemünde.

Many months passed before reports started coming in to British Intelligence about long-range rockets being developed at Peenemünde. In the Oslo Report, the document that had been left on the window sill of the British Consulate in Oslo in November 1939, there had been references to the development of rocket technology. Now this suddenly seemed far more threatening and real. In April 1943, the Joint Intelligence Committee and the Chiefs of Staff thought that Churchill should be made aware of the intelligence reports. Churchill agreed to establish a group to investigate and assess the threat. Duncan Sandys, his son-in-law and an expert on weapons development, was put in charge of the investigation, code-named 'Operation Crossbow'.

Sandys and his team visited Medmenham and studied the aerial photographs, and went through the intelligence reports, some of which had come from the Polish Resistance. But none of it seemed to make much sense and there was an intense debate as to what the German scientists were up to. Then, in late June, a Mosquito photo reconnaissance aircraft made another pass across Peenemünde at high altitude. The day was clear and sunny and when the photographs were developed they revealed what seemed to be the answer to the mystery. The Mosquito had managed to photograph two rockets lying on their transporters alongside a tower structure. By measuring the shadows and knowing the exact time of day the photos had been taken, the interpreters were able to calculate that the rockets were thirty-eight feet long and that they were being assembled inside some sort of launch site.

On the evening of 29 June, Churchill chaired a high-level meeting of the Cabinet Defence Committee. Sandys and his team reported their fears that a form of long-range missile was being constructed. General Brooke, who was present, noted in his diary: 'Arrived at conclusion that definite threat exists.'[1] However, Lord Cherwell, who was also in attendance, was sceptical. He fiercely disputed that any sort of liquid fuel could have been developed as a propellant, and said that no rocket could be made to carry a sufficient payload to cause serious damage. On both these points he was proved entirely wrong. He argued that the Germans were probably developing some form of jet-propelled pilotless flying bomb. In this he was to be proved right. What the scientists in London did not know for sure was that the Germans were experimenting with the production of *two* new types of bomb at Peenemünde. The first, made by a team headed by the brilliant German rocket scientist Werner von Braun, was the A-4 ballistic missile, a rocket which could carry a one-ton payload over a distance of 90–130 miles at over 2000 m.p.h. Several of these missiles had already been successfully launched. Meanwhile, General Dornberger of the German Army, who was running the research station, was also developing a prototype cruise missile for the Luftwaffe, which was launched from a hundred-yard-long ramp.

For some time, the British research teams would be confused by the fact that the Germans were developing these two separate and distinct technologies simultaneously. There was considerable difference of opinion between the British scientists as to what was happening at Peenemünde, and as ever with Cherwell this soon became personal in his sharp criticism of Sandys. But at the key meeting on 29 June, Churchill sided with Sandys rather than Cherwell and authorised the heavy bombing of Peenemünde along with detailed surveillance of all of northern France within 130 miles of London to try to find launch sites for the secret weapons. The group agreed that London was the most likely target, and plans were laid in utmost secrecy for the mass evacuation of children and pregnant women in the event of the use of the terror weapons. It must have seemed that just as one threat to Britain's survival from the U-boats in the Atlantic had been defeated, another, potentially even more deadly threat had appeared.

In June, Hitler himself visited Peenemünde to inspect the progress of the experimental rocket research. He was delighted by what he saw. He believed that the use of these new weapons could turn the course of the war and he told his military chiefs that London would be flattened and Britain forced to capitulate. The date on which the attacks would begin was set for 20 October 1943. Hitler said that tens of thousands of the missiles would be used.

The debate on the scale of the threat continued in London with widespread divergence of views. The home security people believed the warheads could contain ten or even twenty tons of high explosive and were fearful of catastrophic damage to the capital. Cherwell continued to argue that nothing on this scale was technically possible. Churchill later wrote that listening to the opposing arguments, 'it might have seemed at times that the two protagonists were divided as to whether the attack by the self propelled weapons would be annihilating or comparatively unimportant'.[2]

On the night of 17 August, 571 heavy bombers hit Peenemünde. The bombing caused widespread destruction. More than seven hundred scientists and workers were killed, including Dornberger's deputy, who had designed the engine for the A-4 rocket. Forty of the bombers were shot down by enemy fighters. It was a high rate of loss, but the German research had been set back by several critical months. Furthermore, the Germans now decided to relocate this scientific work. Some of it was transferred to underground factories in the Harz Mountains. The work on the A-4 was relocated to an SS artillery range near Blizna in Poland. But it was the flying bombs, the prototype cruise missiles, that would be used first against Britain.

Assisted by reports coming in from the French Resistance, the scientists worked out what the launch sites for these new flying bombs looked like. Each site contained three buildings, the first about 260 feet long and 10 feet wide. It was here that the German engineers constructed the ramp from which the missile was fired. The other two buildings were smaller and always occupied the same relative positions to the long, thin building. In them the missiles were assembled, armed and fuelled. After scouring aerial photographs of the northern French coast, dozens of potential sites were identified by the photo interpreters at Medmenham. In each one, the ramp was being lined up to point exactly in the direction of London. A total of ninety-six of these sites were then bombed and destroyed. All of this further delayed the start of the flying bomb offensive. The first attack was made a week after D-Day, on 13 June 1944, nearly eight months after Hitler had said the missiles would be launched. Goebbels, the Nazi Propaganda Minister, named the missiles '*Vergeltungswaffe*' – literally 'retaliation' or 'vengeance weapons' for the invasion. The flying bomb became known in Britain as the 'V-1', with the 'V' standing for 'vengeance'. Eisenhower wrote later that if the Germans had got the V-1s operating six months earlier, as Hitler had planned, then they could have caused havoc with the preparations for D-Day. He even went as far as to say: '"Overlord" might have been written off.'[3]

As it was, Londoners now experienced a second Blitz. In the next five weeks about three thousand V-1s were fired at the capital. It was a terrible experience to be under these flying bombs. Londoners heard the buzz of their engines as they flew over, hence their popular names 'buzz bombs' and 'doodle-bugs'. Next the engine would cut out and everything fell silent for a few seconds. Then the bomb crashed to the ground, causing a huge explosion. The suspense was appalling. The flying bombs could come over at day or night and regardless of the weather. They were impersonal, indiscriminate killers and made people feel helpless as there was nothing they could do to defend themselves except run for the shelters if there were enough time.

At first it seemed as if there were little defence against the flying bombs once they had been launched. They were difficult to shoot down from the air because of their speed and small size. If an interceptor fighter aircraft got close enough to shoot at one, the fighter would probably be destroyed in the massive explosion that followed. The fastest RAF fighters, Spitfires and Tempests, stripped down to add a few extra miles per hour to their speed, were just about able to catch the jet-propelled flying bombs out over the Channel, and occasionally a brave pilot was able to get close enough to tip one with his wing, so as to deflect it off course and send it crashing into the sea. More effective was a new form of radar known as 'SCR-584' that was developed by scientists at the Radlab, a specialised unit at the Massachusetts Institute of Technology in America. This system could track a flying bomb and fire anti-aircraft shells automatically, bypassing manual operation of the guns. Churchill personally intervened to get this system shipped from America in significant numbers. Combining SCR-584 with new proximity fuses that ignited when the anti-aircraft shells were near to the bomb was the most effective way of striking at the V-1s. Another ingenious way of reducing their impact was by letting it be known that many of the bombs were overshooting London and landing to the north of the city. This piece of disinformation resulted in the Germans resetting the controls of the V-1s to crash to land earlier. This meant that hundreds of flying bombs landed harmlessly in the fields of Sussex or Kent and never reached London.

By the end of August, only one in seven of the flying bombs fired against London was getting through. Of the total 8500 bombs launched against London, a total of 2400 landed somewhere in the South-East. By September, the threat from the V-1s had virtually come to an end as the advancing armies had overrun most of the launch sites in northern France. However, 24,000 civilians were killed or seriously injured and 750,000 homes were damaged during the V-1 flying bomb offensive.

However, just as one threat faded, the second V-weapon began to land on London. These rocket missiles were propelled by the combustion of alcohol and liquid oxygen. Controlled by gyroscopes, they flew up to a height of about fifty miles before heading back to earth in a huge parabola. They had a range of about two hundred miles and the entire journey took just three or four minutes. Their one-ton warhead was about the same as that fitted to the V-1. After the plot on his life on 20 July 1944, Hitler put Himmler and the SS in charge of all his special weapons projects. The SS rushed through the readying of the rockets and the first one landed on Chiswick in west London on the evening of 8 September. There was absolutely no defence against this silent and deadly weapon, known as the 'V-2', other than to destroy the launch sites. As the Allied armies advanced through France and Belgium, most of the launch pads were moved to Holland, many near The Hague. The Germans thought the Allies would not want to bomb the Dutch capital. However, British and American bombers constantly harassed the known launch sites and the missile production centres. This reduced the number of missiles being produced from the intended number of nine hundred per month to roughly half of this. In total, over thirteen hundred V-2 rockets were fired on England and a slightly larger number were fired on Belgium, mostly on the giant port of Antwerp after the Allied armies had captured it. About five hundred reached London and over nine thousand civilians were killed or seriously injured. The V-2 threat was not finally lifted until the last launch sites were captured in March 1945.

The debate about the effectiveness of the V-weapons intensified after the war when Albert Speer, the German Minister of Munitions, said that they had been a massive waste of resources. Speer argued that if all the expertise and materials that had gone into flying bomb and rocket production had instead gone into the development of more fighter aircraft then the bombing offensive against Germany might have been defeated. In this sense, Hitler's obsession with his secret weapons that he thought could yet win the war was a stroke of luck for the Allies. But this new rocket missile technology was clearly a harbinger of things to come, and as the war came near to its end the Americans and Soviets raced to capture as much of the German rocket technology as possible. The United States launched Operation Paperclip, in which they rounded up many German scientists. Werner von Braun and his team preferred to surrender to the Americans rather than the Russians, and under Paperclip they were taken to White Sands in New Mexico, where a new US rocket research establishment was created. In the decades that followed, von Braun and his brilliant engineers played a key role in the American space programme. Twenty-seven years after the first ballistic missile had been launched from Peenemünde, *Apollo 11* astronauts walked on the moon.

With the war in its final phases, and with the Americans and Soviets taking on the major roles in defeating Hitler in Europe and the Japanese in the Pacific, Churchill's status clearly diminished. But there were still major disputes with his military commanders over how best to use the resources they had. The biggest disagreement arose over plans for the deployment of Britain's armed forces in the Far East. At the heart of the argument between Churchill and his Chiefs of Staff were two different approaches as to how best to defeat the Japanese. Churchill wanted to keep the core of the British effort in the Indian Ocean, directed at amphibious operations against Sumatra and further east towards Malaya and Singapore. The Chiefs of Staff wanted the limited British war effort in the Pacific to be lined up alongside the Americans. They produced powerful arguments to show that victory would come sooner if the ships of the Royal Navy, the aircraft of the RAF and the soldiers of the British Army were assembled in Australasia to fight along the left of the main American advance in the South-West Pacific and ultimately towards Japan. Churchill and the War Cabinet took a political view that British interests would be best served if British forces liberated the territories that had been lost in 1941 and 1942. This would help re-establish British power in the region as it had been before the war. The Chiefs of Staff took a more pragmatic view on how best to strike at the Japanese, recognising Britain's junior relationship in this American-dominated theatre. Churchill, as so often in his life, was more interested in maintaining British power and prestige in India than, in this case, advancing through lands he had barely heard of.

The dispute grew to a head in March 1944, three months before Overlord. In his diary, Brooke described several difficult and tense meetings with Churchill. After one such he wrote: 'Now that I know him well episodes such as Antwerp and the Dardanelles no longer puzzle me. But meanwhile I often doubt whether I am going mad or he is really sane.' Then, a few days later, he wrote of 'heated discussions' with Churchill and of a 'desperate meeting'. He was clearly at the end of his tether. Interestingly, later that day, Churchill invited Brooke to dinner. The CIGS thought he was going to be sacked for his outspoken opposition. Instead, Churchill was quite charming and clearly wanted to make up for some of the 'rough passages of the day'.

But the disagreement was a deep one and would not go away. After one Chiefs of Staff meeting, all the secretaries and minute-takers were asked to leave the room and the leaders of Britain's military machine discussed between themselves the possibility of a group resignation. Again Brooke wrote: 'I am shattered by the present condition of the PM. He has lost all balance and is in a very dangerous mood.'[4] General Ismay, the linchpin between Churchill and his War Cabinet, on the one hand, and his Chiefs of Staff, on the other, felt impelled to write to the Prime Minister that the division of opinion was so great there was talk of mass resignation. In Ismay's undoubtedly correct view, 'A breach of this kind, undesirable at any time, would be little short of cata-strophic at the present juncture' so close to D-Day.[5]

The argument seems an arcane one now, and it is difficult to imagine why the passions aroused should have been so intense. No doubt it was partly because of the general state of exhausted tension and stress felt by Churchill and his Chiefs of Staff. The campaign in Italy was going badly and the run-up to D-Day made this one of the most decisive moments in the war. And at the time, of course, no one could know that the war in the Pacific would end so soon after the end of the war in Europe. It was generally felt that the conflict against Japan would con-tinue at least into the summer of 1946 and that strategic decisions affecting that theatre were of vital significance. But it is difficult now to have much sympathy for either side in the dispute. Churchill, his Cabinet and the Foreign Office had all failed to spot the huge changes that had taken place in South-East Asia, where powerful anti-colonial movements had been unleashed by the Japanese successes. The prospect of restoring Britain's pre-war authority in the region was pure chimera. And the idea of landing troops in Sumatra now seems absurd. On the other hand, the soldier's job was to do whatever his polit-ical chiefs demanded, and Brooke's rigid hostility seems out of proportion and character. Even his supporters have not defended him in this quarrel.[6]

Churchill did as he often did at times of crisis and appealed to Roosevelt, who responded by telling him to concentrate on the Indian Ocean. After the Battles of Midway and the Coral Sea, the US Navy was already in command of much of the vastness of the Pacific, and later in the year it would win a decisive victory against the Imperial Japanese Navy at the Battle of Leyte Gulf in the Philippines. Churchill felt his case had been strengthened by the President's view and he wrote a paper outlining his position on future Pacific strategy that he asked the Chiefs of Staff to approve. Brooke wrote: 'We cannot accept it as it stands, and it would be better if we all three resigned rather than accept his solution.'[7] The crisis was defused only when the US Joint Chiefs of Staff announced there were no plans for a major amphibious operation in South-East Asia that year. Both Churchill's grand plan for Sumatra and the Chiefs of Staff's pro-American strategy were redundant. General William Slim would go on to win a considerable victory over Japanese land forces in Burma, so by the time the Pacific question next entered the frame it would be in completely different circumstances.

The crisis marks the lowest point in Churchill's relationship with Brooke and the Chiefs of Staff. Churchill was utterly exhausted by the unrelenting struggle of the war. After one meeting, Brooke wrote: 'He kept yawning and saying he felt desperately tired.' After another: 'PM aged, tired and failing to really grasp matters. It is a depressing sight to see him gradually deteriorating. I wonder how long he will last, not long enough to see the war through I fear.'[8]

But Churchill did recover his strength and his spirits, as he had done before. Only a month later, after dinner at Chequers, he took Brooke aside and told him how much he valued him. Brooke later reflected:

Considering the difficult times I had had recently with Winston I appreciated tremendously his kindness in passing on these remarks to me . . . He was an outstanding mixture, could drive you to complete desperation and to the brink of despair for weeks on end, and then would ask you to spend a couple of hours or so alone with him and would produce the most homely and attractive personality. All that unrelenting tension was temporarily relaxed . . . and you left him with the feeling that you would do anything within your power to help carry the stupendous burden he had shouldered.[9]

Churchill knew he could not push his military chiefs to resignation at this critical moment of the war. A brutal confrontation had been avoided and his charm had once again worked, for now.

The triumph of D-Day and the successful landing of men and *matériel* in Normandy were great boosts to Churchill and his Chiefs of Staff. One of the most complex and ambitious operations of the war had been a success and, despite a heavy storm in mid-June, Port Winston at Arromanches continued to operate as a floating harbour, receiving vast numbers of men and huge quantities of supplies. But the good news of the landings was soon followed by disappointing news as the Allies failed to extend their bridgehead in the way that had been hoped. Monty's troops took until the second week of July to capture the French city of Caen, only ten miles inland. Monty claimed his strategy was to draw the German armour to the east, around Caen, and so allow the Americans to break out in the west. After a slow start, the Germans sent heavy reinforcements to this front, including two crack SS panzer divisions from the Eastern Front, and tried repeatedly to split the bridgehead and throw the Allies back into the sea. Both sides refused to give an inch. The high, thick hedgerows that were common in Normandy made ideal conditions for defenders. By the end of June, one million soldiers were facing each other in bitter fighting. Monty tried to use his infantry to break through in an assault code-named 'Epsom' and then his armour in an attack code-named 'Goodwood'. But each attempt brought a tougher German defensive response. By mid-July, a total of eight panzer divisions, six of them elite SS units, were facing Monty's British and Canadian troops around Caen. It became clear to many that the German tanks, especially the massive sixty-ton Tiger, were far superior to the Allied armour. And the German artillery, particularly the much-feared 88mm gun, which was used as an anti-tank weapon, was superior to any artillery piece in the Allied arsenal.

In the circumstances, it might have been expected that Churchill would give vent to his frustrations at the slowness of the offensive, as he had done earlier in the war with Wavell and Auchinleck, by demanding action and pestering his field commanders with instructions and briefs. But Brooke managed to keep Churchill at arm's length from Monty, who was left to carry on with his operational planning unhindered by interventions from the Prime Minister. Churchill was also no doubt acutely aware of Monty's popularity with the public, which was at a different level to that of his previous generals. Unfortunately, no breakout came and it was Eisenhower who grew increasingly frustrated by Monty's lack of progress. The tension between the two commanders grew from this point and lasted right on to the battle of the memoirs and reputations in the post-war period.

When the breakout did finally take place in the last week of July, the American First Army commanded by General Bradley led the way in Operation Cobra. After a heavy air bombardment, the US VII Corps launched the assault near St Lô.

Monty's strategy of tying down the German armour in the east now paid off as there were still fourteen German divisions gathered around Caen, leaving only eleven weakened divisions facing fifteen American divisions in the west. Once the German line began to crumble, the American Sherman tanks were ordered in and their crews began to liberate town after town as they sped onwards. On 1 August, the US Army reorganised in Normandy and Bradley brought in General Patton to lead the newly created 3rd Army. Renowned for his aggressive spirit, Patton brought a new dynamism just when the breakout was gathering speed. A key bridge was captured intact at Pontauban, and Patton's men raced across it and fanned out into Brittany. Some of them advanced fifty miles in four days. Hitler ordered his commanders to give no ground, as he had done at Stalingrad, and replaced some of his senior generals. Rommel, who had done so much to hold up the Allied advance in Normandy, was severely wounded by an RAF attack upon his staff car. His career was over. Restrained by lack of fuel and ammunition, constantly harried from the air, and now prevented from making tactical withdrawals to reorganise in strength, the German troops still fought on. Especially tenacious were the fanatical SS panzer grenadiers whose ferocious support for their Fuehrer was only hardened after news got out of the unsuccessful bomb plot against his life. But no army, no matter how determined, could triumph against the overwhelming odds the Germans now faced. Every tank they lost was a permanent loss. For the Allies, every tank lost was replaced by two or three more within days.

Hitler, suspicious of all his army commanders after the plot against him, ordered General von Kluge, Rommel's replacement, to counter-attack. This he could do only reluctantly by moving his armour westwards. The counter-attack, inevitably, failed and now von Kluge's armour was deep inside an Allied

encirclement as Patton's tanks raced eastwards towards the
Loire and the Seine. The Canadians then broke through and
advanced south while Patton turned north. The remnants of the
German Army were surrounded. Some got away through the
Falaise 'gap', but the bulk of Hitler's Army Group West was
destroyed. Like the British Army at Dunkirk, the Germans had
to abandon a lot of their heavy equipment. Ten thousand
German soldiers had been killed and fifty thousand were taken
prisoner. More than two thousand panzers had been lost.
Twenty-seven divisions had ceased to exist. Von Kluge wrote a
note to Hitler saying the war was as good as lost. Then, knowing he would be blamed for the defeat, he committed suicide.
Four days after the collapse at Falaise, on 25 August, the
German garrison in Paris surrendered. It was the culmination
of a huge Allied victory.

Typically, Churchill seized every opportunity to visit the battle-
front. He spent some time in Normandy in July and August,
observing for himself the landscape of war, and eagerly followed the news of the breakout. He was concerned by the
growing American hostility towards Monty. And there were still
arguments over the planned landings in the south of France,
now called 'Operation Dragoon'. Churchill did not want troops
taken from the advance in northern Italy for a mission that he
regarded as entirely unnecessary. Of 250,000 men in the 5th
Army under General Alexander in Italy, roughly 100,000 would
be taken away for Dragoon. Churchill knew that the advance
through Italy and then the plan to cross the Alps and head
towards Vienna would be slowed right down by diverting
troops to the Côte d'Azur. This time his Chiefs of Staff agreed
totally with him. Churchill appealed to Roosevelt to cancel the
operation. The President again refused Churchill's plea and
stood by his Joint Chiefs, who maintained their faith in the landings. On 5 August, Churchill spent more than six hours at

Eisenhower's headquarters in Portsmouth, trying to persuade him to cancel the operation. Eisenhower's aide-de-camp, Captain Harry Butcher, recalled that 'Ike said "No", continued saying "No" all afternoon, and ended saying "No" in every form of the English language at his command.'[10]

Six days later, Churchill flew to Italy to see the situation there for himself. He met with Marshal Tito, the leader of the Partisans in Yugoslavia, who had maintained an epic and brilliant struggle against the German occupying forces, supported by SOE. Churchill found Tito a curious but impressive figure and was pleased to hear that he was not committed to a communist future for Yugoslavia. Churchill then went on to witness the landings in southern France. He watched the shore bombardment from the British destroyer HMS *Kimberley*. For Churchill, the event was the great anticlimax he had predicted, and he noted that 'not a shot was fired either at the approaching flotillas or on the beaches'. His return was so uneventful that he even borrowed a novel from the ship's captain that he sat down and read. He wrote to Clementine that it was one of the best he had read in years.[11] Churchill travelled on to Naples, where he combined talks about the future of Italy with several swimming expeditions along the coast. The Mediterranean sunshine and the bathing refreshed his spirits. He returned to London rested and in good health.

Once again in the war, the victory in Normandy raised the question: what next? How should the Allies pursue their drive eastwards in what they hoped would be the final operation of the war? There were two views on this. Montgomery wanted to go for a single thrust of some forty divisions to strike forwards by the Ardennes, through the Ruhr and across the plains of northern Germany to Berlin. He wanted to lead this himself and so go down in history as the Allied general who won the war in Europe. Eisenhower, on the other hand, favoured a broader, two-

pronged assault upon Germany, with Monty driving his forces
to the north and a southern, American force heading below the
Ardennes and crossing into Germany along the Saar River. This
dispute between the two commanders, who were already not on
the best of terms, grew in intensity over the weeks. Monty did
not like being overruled. But Eisenhower was in command and
the Americans were calling the shots. They would not accept that
a British general could command US troops in such a key operation. Neither Churchill nor his Chiefs of Staff could do anything
to prevent Eisenhower from getting his way.

In September, Monty launched his own highly imaginative
thrust, code-named 'Operation Market Garden'. He intended to
use a sequence of airborne drops to seize a series of key bridges
across Holland right up to the Rhine. A relief column would
then speed across country to link all the captured bridges and
propel the Allies right into Germany itself. All the remaining
V-2 launch sites would be captured and new ports would be
opened up to the Allies to supply the advancing armies. It was
an ambitious attempt to bring the war to a speedy end. Much
of the plan was brilliantly executed. American paratroopers
captured bridges at Eindhoven, Grave and Nijmegen. Unfortunately, alongside the last bridge to be seized by the British 1st
Airborne at Arnhem, the Germans had located a rest camp for
two SS panzer divisions that had been pulled out of the line.
Photo reconnaissance had identified that these elite German
units were in the vicinity, but the British commanders refused
to adapt their plan. The British paratroopers, known as the 'Red
Devils' and commanded by Colonel John Frost, conducted a
heroic defence of the bridge. But the land troops failed to reach
them and this led to the collapse of Monty's daring plan. The
bridge at Arnhem is now remembered as the 'bridge too far'.
With the failure here went all hope of a fast strike into Germany
and the end of the war by Christmas.

In September, Churchill and his Chiefs of Staff travelled again to Quebec to meet Roosevelt and the Joint Chiefs for what proved to be the last Anglo-American summit of the war, code-named 'Octagon'. By this time, most of Churchill's attention was focused not so much on finishing the war, the result of which was by now a foregone conclusion, but on the post-war world and his fears of a dominant Russia. As the Red Army advanced west, beyond the borders of the Soviet Union, the question arose as to who should take over liberated territories. The first instance in which this became acute was in Poland. In the month before Octagon, the Polish Resistance rose up against the German occupation as the Red Army approached Warsaw. Their intention was to overthrow German rule and replace it with an independent Polish government based around the leadership in exile in London. The Germans responded by putting down the uprising with ferocious brutality. The Red Army now paused on the Vistula River, only a few miles from Warsaw, but Stalin ordered his soldiers not to intervene. He wanted his own communist supporters in the Polish Committee of National Liberation to take control, so he was happy to see the Poles who supported the London democrats massacred. Churchill and the Cabinet, along with public opinion in much of the West, were outraged by Stalin's refusal to help the Polish rebels. After all, Britain had gone to war in 1939 to defend Poland, and Churchill felt a particular responsibility for the country. He pleaded with Stalin to intervene. The Soviet leader said this was impossible – his troops needed time to regroup. There was nothing that British or American forces could do to aid the rebels as Warsaw was too far away to permit direct military intervention. All that could be done was for the Allied air forces to drop in supplies to assist the rebels. But no supply aircraft that could fly to Warsaw had the range to make it back, so they would have to land at a Soviet airfield to refuel and return. Stalin absolutely prohibited this. Churchill was furious. The entire War Cabinet fumed and protested to Stalin. Still, he did nothing. For sixty-three days, the Polish Resistance fighters held off the German onslaught armed only with rifles and small arms. In the end 200,000 Poles were slaughtered by the Nazis in an orgy of violence. Stalin had allowed the free Polish Resistance movement to be destroyed so that he could hand Poland over to his own stooges.

For Churchill, this rang alarm bells about the possibility of a post-war Europe divided between East and West. At Quebec, it was clear that Roosevelt was far less concerned about this. The two leaders, along with the Combined Chiefs of Staff, discussed and agreed the details for the next phase of the war. In Europe, Eisenhower's plan for a two-pronged assault on Germany was endorsed. In Italy, the Americans now raised no objections to an advance across the Alps and into southern Europe. Churchill talked about his plan to liberate Vienna. In the Far East, Churchill wanted to assure the Americans of the British commitment to play its role in the defeat of Japan. He was again thinking about the post-war era and wanted Britain to have earned the right to restore its possessions that had been captured by Japan. He did not want these to be handed back by a peace treaty; he wanted to win them back by force of arms. However, the Americans were deeply suspicious of Britain's imperial intentions in the Pacific. The US military chiefs wanted to see Burma recaptured as this was vital to supplying China. But Admiral King, who had never been much of a friend to Britain, saw the naval war in the Pacific as an exclusively American affair. In the end, the President overruled his own admiral and agreed that the Royal Navy could contribute to the defeat of Japan. But General Marshall pointed out that the United States now had enough heavy bombers to carry out the bombing of Japan. It seemed that once Germany had been defeated, the

RAF's role in the Far East would be limited to operations in Burma and the Indian Ocean. It was estimated that the war against Japan would continue for about eighteen months after the defeat of Germany.

At Octagon, Churchill's relationship with Roosevelt was much cooler than it had been earlier in the war. His friend and intermediary Harry Hopkins, who had done so much to help cement relations between Britain and the United States since 1941, had fallen out with the President and was not present in Quebec. Churchill spent a couple of days with Roosevelt at his home in Hyde Park after the summit, and then left for home.

Once the Octagon meetings had set the war strategy for the next six to twelve months, there was little in reality for Churchill to do as a warlord. It was up to the generals now to defeat Hitler in the West and to continue the advance towards Japan in the Pacific. But Churchill became even more concerned with the growing power of the Soviet Union as the Red Army looked set to liberate more of Eastern Europe. So, in October, he decided to pay another visit to 'Uncle Joe' in Moscow. Roosevelt was facing an election in November and could not travel, but he approved of the visit.

On 9 October, Churchill arrived in Moscow and late that evening he had his first meeting with Stalin. Anthony Eden, the Foreign Secretary, was with him. Molotov, the Soviet Foreign Minister, accompanied Stalin. Apart from the two translators, it was just these four men who met together late at night in the Kremlin. They talked about Poland and about Greece, and Churchill made it clear that he wanted to avoid civil wars in the countries of Europe over who should be in power after the war. The conversation was frank and was going well, so sensing this opening meeting was a good moment for business, Churchill suggested the two leaders try to settle their affairs in the Balkans. He said he had a 'naughty document'. Churchill proposed a simple breakdown of spheres of interest: Romania should be 90 per cent Soviet and 10 per cent British; Greece 90 per cent British-American and 10 per cent Soviet; Yugoslavia and Hungary both 50–50; and Bulgaria 75 per cent Soviet and 25 per cent British-American. While this was being translated, Churchill wrote it out on a piece of paper. There was a pause. Then Stalin reached for a blue pencil and put a big tick on the piece of paper. A long silence followed. Then Churchill said: 'Might it not be thought rather cynical if it seemed we had disposed of these issues, so fateful to millions of people in such an offhand manner? Let us burn the paper.' Stalin replied: 'No, you keep it.'[12]

Churchill was pleased with this exchange, although it led to much haggling between Eden and Molotov in the following days. Churchill was also delighted with the rapport he once again developed with Stalin in the face-to-face meetings that followed over the next ten days. There was agreement on nearly all the military issues. And there were discussions about the possibility of breaking up Germany after the war. Despite Stalin's joke a year before at Teheran about executing the top Nazis, the Soviet leader now firmly supported the idea of public trials of the Nazi leadership after the war was won. Only the issue of Poland continued to divide the two leaders. The pro-Western Polish politicians came to present their case to Churchill and Stalin in Moscow. Then the Soviet-backed team did the same. Churchill and Eden had no time for these pawns of Stalin. They thought, correctly, that the Polish communists were just repeating parrot-like a script that had been written for them by the Soviets. After one speech by the Polish communist leader, Churchill caught Stalin's look and saw 'an understanding twinkle in his expressive eyes, as much as to say, "What about that for our Soviet teaching!"'[13] Despite days of negotiations, the future frontiers and government of Poland were left unresolved.

Nevertheless, in the course of these meetings, Churchill and Stalin built up almost a friendship. More than anything else, they were united in their struggle to defeat Nazism. One evening they attended the Bolshoi Ballet, and when they appeared in the Royal Box together they received a long, rapturous reception from the audience. At several late-night sessions, many toasts were drunk to each other and to the joint interests of the British and Russian people. At one point, some-one described the two leaders along with Roosevelt as the 'Holy Trinity'. Stalin responded by saying: 'If that is so, Churchill must be the Holy Ghost. He flies around so much.'[14] But despite the bonhomie, deep disagreements about the shape of the post-war world divided the two men. And however jovial Stalin might be in Churchill's presence, he would still ruthlessly pursue his own interests when it came to it.

Churchill continued to travel. He spent some time in Cairo on his return from Moscow. In November, he attended a parade down the Champs Elysées in Paris with Charles de Gaulle amid wildly cheering crowds. Meanwhile, the war was not going well. Churchill had refused to believe various intelligence reports that Germany would be defeated by the end of the year. He was right. In Italy, the advance came to a halt for the winter in the Apennine Mountains. In France, in mid-December, Hitler's armies launched one final surprise counter-attack in the Ardennes with ten panzer divisions in what came to be called the 'Battle of the Bulge' as German troops created a huge salient, or bulge, in the Allied line. The intention was to divide the British and American armies and strike at Antwerp. But the Germans never completely broke through. Their tanks were awesome weapons but, desperately short of fuel, the Germans could not afford to run them for long. This, along with a heroic stand by US soldiers at the major crossroads at Bastogne, defeated this last attempt by Hitler's armies to reverse the course of the war. But in the midst of a cold winter, further major campaigning had to be postponed to the spring.

Churchill felt a strong affinity for Greece, as British soldiers had fought and died there trying to defend the country in 1941. When the Germans withdrew their garrison in October 1944, a flying column of SAS troops entered the country and raced to Athens, where they were rapturously received as liberators. But the people of Greece were in a desperate state after four years of brutal German occupation. Many now went over to support ELAS, the communist nationalist movement, despite British attempts to install King George and a government led by Prime Minister George Papandreou. During December, the tensions erupted into civil war in Athens. Churchill ordered the British commander, General Ronald Scobie, to fire on the communists in order to maintain order. This sided Britain with the royalists in the civil war and looked like interference in a foreign state's internal politics. In the United States, it was seen as an attempt by Britain to sustain its power in the region and was widely denounced. Churchill once again felt he had to intervene in person. On the afternoon of 24 December, he decided to fly to Athens that night. At home, Clementine was preparing an eagerly anticipated family Christmas, beginning with a children's party that evening. Despite being so used to last-minute changes of plan, she was deeply upset by her husband's decision to leave. This was one of the few times in the whole war when she burst into 'floods of tears' and was 'laid low' by Winston's stubbornness.[15]

Churchill arrived in Athens on Christmas Day and had to be given an armed escort through the dangerously divided city, where British troops were still engaged in street fighting with the communists. It was decided that he would be safest on board HMS *Ajax* in Piraeus harbour. Here he met Archbishop Damaskinos, the Greek Orthodox Patriarch, a tall, impressive man who had once been a champion wrestler. To Churchill, Damaskinos seemed to have the authority to preside over matters. Churchill suggested calling a meeting of all rival groups on the following day, to be chaired by Damaskinos. The Patriarch agreed. The meeting was held in the Greek Foreign Ministry at 6 p.m. on Boxing Day. It was bitterly cold and there was no heating. A few hurricane lamps cast an eerie glow upon the scene. But all parties, including the communists, turned up for the roundtable talks. After opening the talks, Churchill and the British delegation then withdrew and left the tough negotiating to the Greeks themselves. A couple of days later, it was agreed that Churchill should ask the Greek King, who was in exile in London, to appoint the Archbishop as Regent. When he returned home, Churchill persuaded King George in an all-night session of the wisdom of this course of action. Damaskinos subsequently became Regent and the effective ruler of Greece. A truce was signed with the communists in January. Churchill was pleased that he had helped to save Greece from communist subjugation. He was convinced that, having defeated fascism in Europe, the next threat would be that of communism. In this he was way ahead of most Western leaders and was already anticipating the divisions of the Cold War. He saw Europe now made up of countries that would either fall under the 'heel' of communism or be saved for the West. He was convinced that he had done the right thing for Greece and would still try his best to rescue Poland.

In February 1945, the Big Three met once again, this time at the old Livadia summer palace of the tsars at Yalta in the Crimea. Roosevelt, who had been re-elected President at the end of 1944, was exhausted by the journey. In the group photos taken at Yalta he looks haggard and drawn. Churchill, too, was exhausted by the years and months of relentless pressure. In one sense, the meetings at Yalta were the high water mark of

Allied wartime collaboration. Stalin confirmed that he would join the war against Japan within three months of the defeat of Hitler. The Allied military leaders discussed what support they could give each other, and the Soviets formally requested bombing the cities behind the German lines in the East. This was one of the factors, as we have seen, that led to the destruction of Dresden by British and American heavy bombers. On the other hand, Yalta also represents the beginnings of the Cold War. Stalin was suspicious of Anglo-American plans for the United Nations Organisation that had been drawn up at Dumbarton Oaks in Washington. It was proposed there would be a consultative General Assembly, to which all nations could belong, and a Security Council. This Security Council would have the teeth that the failed League of Nations never had, with authority to order executive action on behalf of the UN, even to the extent of going to war on its behalf. Stalin thought the Soviet Union could easily be outvoted by Britain and America in such a body, and that the status the USSR had won by playing the major role in the defeat of Hitler was not reflected in the UN structure. It was agreed that various Soviet republics could also join the UN, and that each of the great powers would have a veto over Security Council resolutions.

There was still further dispute over the governance of Poland. Churchill again explained that this was a matter of 'honour' for Britain. Stalin explained that it was a matter of 'security' for him. Russia had been attacked twice in the last thirty years through the 'Polish corridor'. It must not be allowed to happen again. By this he clearly meant that he wanted control of the Polish government. And it was 'security' that underpinned Stalin's strategy for the whole of Eastern Europe. Churchill feared that Stalin wanted to create a buffer of satellite states. In the end, Roosevelt and Churchill largely gave in to Stalin's demands for a new border with Poland, which was now moved westwards. In compensation, Poland's border with Germany was shifted further west, into what had been German territory. Stalin agreed to free elections in Poland and he signed a Declaration on Liberated Europe that pledged support for reconstruction based on free elections. For now, the Western leaders took Stalin at his word.

It was agreed to divide Germany into four zones of military occupation: Soviet, American, British and, later, French. There was an argument over reparations. Stalin insisted on massive reparations, partly to compensate for the vast destruction caused by the German armies in the Soviet Union. Here, 32,000 factories were in ruin, 50,000 miles of railway track had been destroyed, 1710 towns had been devastated and about 100,000 collective farms had been burned to the ground. But Stalin also believed in reparations as a form of punishment and as a symbol of a victor's rights. The Western leaders thought that reparations had prevented Germany from recovering after the First World War. They took a more pragmatic view and wanted to restore Germany, not destroy it, after this war. Eventually, a compromise was agreed.

Yalta revealed major fissures in the Grand Alliance that had been held together by the common objective of defeating Hitler. Churchill was increasingly suspicious of Soviet post-war ambitions. Roosevelt, on the other hand, thought that collaboration with the Soviets was the only way of preventing post-war disputes. Churchill was pained by Roosevelt's attitude towards him at Yalta. In chummying-up to Stalin, the President seemed to be distancing himself from his old ally and friend.

After the summit, Churchill passed through Athens. This time, he was able to ride through the streets in an open-top car amid cheering crowds where only a few weeks before vicious street fighting had taken place. He went on to Alexandria in Egypt, where he had lunch with Roosevelt. Churchill noted how frail the President looked. After lunch Churchill said farewell to his friend. It was the last time he would see him.

In March, the Allied armies launched their final attack upon the Third Reich. The Rhine had been a barrier to the invasion of Germany since Roman times. Now it was the last major obstacle facing the Allies. Eisenhower was again in overall command. Monty and the 21st Army Group were to cross the Rhine in the north, between Cleves and Düsseldorf, and head across the German plain to Hamburg and the Baltic. Bradley and the 12th Army Group were to cross the river further south. It was a combined boat, air and land operation, the largest single offensive since D-Day. There were forty bridges across the Rhine. Nazi engineers planned to destroy all of them, but in a lucky break in early March, the US 9th Armored Division captured one bridge intact at Remagen. Tens of thousands of men and hundreds of tanks poured across it. Hitler was so furious when he heard of the failure to blow up the bridge that he sent out Gestapo squads to execute all those responsible. Later in the month, Patton launched an attack across the Rhine at Oppenheim. As in Sicily, Patton was determined to be ahead of Monty, and his men crossed the river one day before those of his British rival. Further north, Monty launched a massive artillery bombardment, followed by attacks with heavy bombers, against the German defenders who were dug in along the eastern bank of the river. On the night of 23 March, British commandos led the crossing at Wessel in armoured amphibious vehicles called 'Buffaloes'.

Churchill, always keen to be an observer at these big military shows, asked to join Monty at his headquarters for the crossing. Brooke, who accompanied him, was not happy. He wrote in his diary: 'All he will do is to endanger his life unnecessarily and to get in everybody's way and be a damned nuisance to everybody. However nothing on earth will stop him!'[16] They arrived

at Monty's forward HQ on the evening of the attack, delighted
to be on German soil. After dinner, Monty retired to bed and
Churchill and Brooke went for a walk in the moonlight. A few
miles away, a furious battle was taking place. The two men took
this moment to look back over the struggles they had been
through together. Churchill told his leading general how much
he appreciated him. It was a moment of personal warmth on
the fringes of the last great battle of the war involving British
soldiers.

On the morning of 24 March, Churchill and Brooke were
escorted to the top of a hill to watch the huge airborne landing
behind German lines at Wessel. Called 'Operation Varsity', this
was the biggest air drop since D-Day. The vast air armada, with
over 1700 transport planes and 1300 gliders, took two hours to
pass. Churchill was as thrilled to watch this huge military operation as he had been to observe the battle at Omdurman, when
the British Army attacked at Khartoum, nearly fifty years
before. The little boy in him who had enjoyed playing with his
toy soldiers was still there. War was still a romantic and heroic
escapade for Churchill, despite everything he had lived
through. The following day, Churchill and Brooke even managed to cross the Rhine in a landing craft and stood for a few
minutes on the eastern bank, examining the German defences.
Back on the western bank, they scrambled about on the remains
of the bridge at Wessel when they came under sniper fire. Then
some shelling opened up near them. General Simpson, the local
American commander, said he could no longer take responsibility for Churchill's presence and ordered him to leave the
battlefield. Brooke remembered that 'The look on Winston's
face was just like that of a small boy being called away from his
sandcastles on the beach by his nurse! . . . Thank heaven he
came away quietly, it was a sad wrench for him, he was enjoying himself immensely.'[17]

The Rhine crossing was a great military success. Within days, American and British troops, supported now by French soldiers, were striking hard into Germany. Eisenhower's northern and southern thrusts joined up, having encircled the Ruhr and its powerful defences. The German western front was collapsing.

Churchill's position in the spring of 1945 was difficult. He still had the personal prestige and status of a giant, the man who had led Britain from the abyss and was now one of the Big Three. But Britain was very much the junior partner in the alliance. The Soviet Union now had the largest army in history, and its enormous suffering in the war, estimated recently at about twenty-seven million dead, left Stalin with a legitimate claim to superpower status in the post-war world. Meanwhile, the United States was ending the war as the greatest industrial power in history. Its economy had more than doubled during the war. Its factories had become the 'arsenal of victory', producing the guns, ships, planes and tanks that had won the war. Nearly 50 per cent of all the world's goods were manufactured in America. There was no question that the United States would be a post-war superpower. Britain, however, was technically bankrupt. The only way it could have authority in world affairs after the war was as an ally of the United States. Churchill had to persuade Roosevelt that the new enemy was going to be the Soviet Union, and steps had to be taken to stand up to Stalin in what the Prime Minister saw as the inevitable East–West divide after the war. However, Roosevelt saw the future differently. He thought he could get along with Stalin and that the United Nations would be the agency of peace. In the end, Churchill's view was proved right. He had correctly foreseen the tensions of the imminent Cold War. Roosevelt's vision died with him.

Over the next month, Churchill continued to be agitated by the Soviet attitude to the end of the war in Europe and by the

lack of progress on Poland's future. It was clear to him that the
Soviets had no intention of implementing free elections there,
as had been agreed at Yalta. Churchill wanted to put pressure
on Stalin and tried to persuade Roosevelt to side with him. But
the President did not want to antagonise Stalin and suggested
a truce. There was a further dispute when the Soviets accused
the Western Allies of negotiating a separate peace treaty with
Germany. Roosevelt worked hard to resolve these issues. On 12
April, he sent a telegram to Churchill, saying: 'I would min-
imise the general Soviet problem as much as possible.'[18] A few
hours after sending this message, Roosevelt collapsed. He
never regained consciousness and died a few hours later. When
he heard the news, Churchill felt he had been struck a physical
blow. Although their views had diverged over recent months,
Roosevelt had been the great ally who had supported Britain's
war effort from the beginning. They had established an almost
daily correspondence. They had enjoyed nine separate meetings and had spent about 120 days in close personal contact.

But Churchill decided not to attend Roosevelt's funeral. This
was particularly surprising bearing in mind his enthusiasm for
getting on a plane and travelling almost anywhere. It was even
more remarkable bearing in mind the opportunity it offered to
meet with the new President and to try to influence his thinking. Churchill claimed he was under pressure to remain in
London during these last days of the war. But that had never
stopped him before. The only conclusion to be drawn is that he
felt seriously let down by the President over recent months.
Foreign Minister Eden attended the funeral on his behalf.

In accordance with the American constitution, the Vice-
President, Harry S Truman, immediately succeeded to the
presidency. 'I feel like I've been struck by a bolt of lightning,' he
told a colleague. Despite his ill health, Roosevelt had kept
Truman woefully in the dark about key strategic and political

developments in the war. Roosevelt had only two private meetings with his Vice-President in the five months since the
election. Now, Truman had to go on a crash course in foreign
and military affairs. Interestingly, he soon took up Churchill's
view, believing that reconciliation with the Soviets was impossible because of their bully-boy tactics, and within months of
taking office he became forcefully anti-Soviet.

In the meantime, for a variety of reasons, Eisenhower had
decided not to try to capture the German capital. First, Berlin
was well inside the Soviet zone of military occupation as agreed
at Yalta, and it was more appropriate for the Red Army to capture the city that was only thirty-five miles from their front line.
Second, he did not see Berlin as a major objective. He wanted
to smash through to the centre of Germany, where he believed
the Nazi government was planning to move. And finally,
despite the German collapse, Eisenhower guessed that there
would still be a ferocious battle for the capital of Hitler's Reich.
So he approached Stalin directly and suggested Allied troops
should halt on the river Elbe. Churchill was deeply opposed to
this and wanted Western troops to seize the prize of Berlin. He
thought this would leave the West in a stronger position in any
post-war conflict. But Marshall and the American Joint Chiefs
backed their field commander. No matter how much the Brits
protested, they were going to finish off the war their way. And
military rather than political priorities prevailed.[19]

It was Stalin who ordered the last assault on Hitler's capital. On 16 April, the final offensive began. Two and a half
million men in the Red Army had assembled along the banks
of the Oder and Neisse rivers. Facing them were 700,000
German troops. Many of them were poorly equipped. They
included boys from the Hitler Youth and old men fighting
to defend Hitler's capital from what they saw as the Bolshevik
hordes. With them were the last Nazi die-hards. They all

fought with a grim fanaticism. Stalin set off his two leading commanders, Marshals Zhukov and Koniev, in a race to battle their way into Berlin and win the accolade of taking the final German surrender.

On the morning of 20 April, the Allies launched their last 'thousand-bomber' raid against Berlin. When it was over, Hitler emerged from his bunker one last time to give medals to the defenders of the city. His hopes were placed in the fanatical determination of teenage boys. But the newsreel cameras caught the desperate twitch in his hands. The two huge Soviet armies advanced into the suburbs of Berlin, district by district, then street by street, then house by house. The fighting was as intense as Eisenhower had feared. The Soviet tanks that had led the Red Army fifteen hundred miles from Stalingrad to Berlin were no good in the urban environment and could be taken out at close range with primitive, easy-to-use weapons. German civilians hid in their cellars. SS execution squads roamed the city in search of deserters. Hundreds of thousands of German women were raped by Soviet soldiers as they advanced through the city. Hitler and his entourage in their bunker under the Reich Chancellery still believed there were armies waiting to rescue them. There were none. As the Red Army finally closed in on the centre of the city, Hitler at last accepted defeat. On the afternoon of 30 April, he shot himself. Churchill was having dinner when his private secretary, John Colville, brought him the news. According to German radio, Hitler had died 'fighting with his last breath against Bolshevism'. Churchill commented dryly: 'Well, I must say, I think he was perfectly right to die like that.'[20]

The fighting went on for a few days longer. The Red Flag was suspended from the top of the Reichstag, the symbol of Hitler's capital, in time for the great socialist parade of May Day. Three hundred thousand Russian soldiers lost their lives in the Battle of Berlin. A few days later, Hitler's generals tried to negotiate terms. They were told that only unconditional surrender of all German armies on every front was acceptable.

On 2 May, German forces in Italy surrendered. On 4 May, Montgomery, at his headquarters on Luneburg Heath, received the surrender of all German forces in north-western Germany, Denmark and Holland. That evening, Churchill called the Chiefs of Staff to the Cabinet Room in Downing Street and thanked them for all they had done. Then he shook their hands, one by one. Brooke noted he had tears in his eyes.[21] On 7 May, the German High Command finally signed a document of unconditional surrender. The guns at last fell silent. The war in Europe was over. Late that evening, Churchill called in Elizabeth Layton, one of his secretaries, to type up his dictation of the speech he would broadcast. 'Hullo, Miss Layton,' he said, 'well the war's over, you've played your part.'[22]

The following day, Tuesday 8 May, was proclaimed Victory in Europe or VE Day. Huge crowds gathered in London. At 3 p.m., Churchill broadcast the news of the surrender to the people of Britain and around the world. 'We may allow ourselves a brief period of rejoicing,' he said. 'But let us not forget for a moment the toil and efforts that lie ahead. Japan, with all her treachery and greed, remains unsubdued.' He ended the short broadcast with the words: 'We must now devote all our strength and resources to the completion of our task, both at home and abroad. Advance, Britannia! Long live the cause of freedom! God save the King.' On the words 'Advance, Britannia', Churchill's voice broke with emotion. That evening, he appeared on a balcony overlooking Whitehall, where vast crowds had assembled. The cheering was intense. 'God bless you all,' he said to the people below, 'this is your victory.' The crowd roared back: 'No – it's yours.'[23] Churchill also appeared with the King and Queen on the balcony of Buckingham Palace. The huge crowds once again cheered rapturously. It was almost

five years to the day since Churchill had been appointed Prime
Minister.

Churchill was still severely agitated by Soviet actions in the
territories they had liberated, or captured, at the end of the war.
On 12 May, he sent a message to President Truman using a
phrase he would later make famous. He wrote: 'An iron curtain
is drawn upon their [the Russian] front. We do not know what
is going on behind.'[24] He feared that Stalin was imposing his
own puppet governments in all the states of Eastern Europe. A
conference was called for the victorious powers. They agreed to
meet in Potsdam, outside devastated Berlin, in July.

Meanwhile, it was clear that the days were numbered for the
coalition government Churchill had led for five years. The
Labour Party took the view that, with Germany defeated, it was
now time for a general election. After all, with all-party agreement, there had not been one for nearly ten years. Churchill
himself argued that Japan should be defeated first, and nothing
should detract from this major objective. However, everyone
still expected that this would take another eighteen months.
Attlee decided that the country could not wait that long and
withdrew the Labour Party from the government. So, on 23 May,
Churchill tendered his resignation to the King. He agreed to
form a caretaker government until the election could be held
on 5 July and all the votes from soldiers serving overseas
counted. With the war in the Far East still raging, the campaigning began.

So great had been the burden of running Britain's war
machine that Churchill had almost completely ignored domestic politics during his five years in office. He had been happy to
leave the Home Front to Attlee, his deputy. And Churchill had
not wanted to make promises about conditions after the war
that he felt he could not deliver. He was very conscious that the
promise to build 'Homes fit for Heroes' after the First World

War had backfired upon Lloyd George. So Churchill had completely failed to pick up on the sea change that had taken place
in the thinking of the British people. He was revered as a war
leader but in conventional politics he was seen as hopelessly
old-fashioned. The country had made a major shift to the left.
There was a strong feeling that, after all people had been through,
there could be no return to the depression and misery of the
1930s. Men and women across Britain wanted full employment,
better housing, national healthcare and social security. Many
had thought long and hard about the future, even if Churchill
had not. The Beveridge Report, which called for a welfare state
to look after British citizens 'from the cradle to the grave', had
sold 635,000 copies. Penguin Specials had debated every aspect
of the future shape of Britain and had sold millions. Even inside
the military, groups like the Army Bureau of Current Affairs
(ABCA) had encouraged political debate about the post-war
world and here, as elsewhere, a broad socialist perspective often
prevailed. Churchill had been opposed to ABCA on the grounds
that it would be bad for military discipline.[25] But now he looked
old, tired and a leftover from an older Britain as he took up
the mantle of leading the Conservative Party in the election
campaign. Also, a remark in a party political broadcast that
likened Labour Party tactics to those of the Gestapo in seeking
to introduce a socialist state was very ill-judged and did him
great harm.

Polling duly took place in early July, but because of the time
it would take to collect and count the three million votes from
army, navy and air force personnel around the world, the result
would not be announced until the end of the month. Meanwhile,
on the 15th, Churchill travelled to Potsdam for the end-of-war
Big Three meeting. The next morning, he met President Truman
for the first time and was impressed, even though Truman was
still desperately new to the complexities of inter-power politics.

Accompanying the novice President was his Secretary of State, James Byrnes, who had been sworn in only three days before leaving for Europe. Before the formal sessions began, Churchill toured the ruins of Berlin. Outside the Chancellery building a small crowd of Berliners gathered. To Churchill's surprise, they cheered him. Already allegiances were changing.

When the conference began, Stalin again pledged his commitment to join the war against Japan. And agreement was swiftly reached on the military occupation of Germany in four separate zones. But there was still disagreement over the Polish borders and reparations. Then, on the 17th, extraordinary news arrived. The Americans had successfully carried out the first atomic bomb test in the New Mexico desert. Reports said the explosion was 'brighter than a thousand suns'. Truman and Churchill were told and immediately realised that this new weapon of war entirely transformed the struggle against Japan. US military planning was for an invasion of Japan some time between November 1945 and spring 1946. Nearly two million men would be involved. General Marshall feared that casualties would be extremely high. On Iwo Jima and Okinawa, Japanese soldiers had fought almost literally to the last bullet. And many had chosen suicide over surrender. Churchill was convinced that Truman would use the atom bomb and the war would be over in a matter of weeks. This meant that they no longer needed the Soviets to join in the war against Japan. The question was, how were they to explain this to Stalin?

Truman decided to tell the Soviet leader about the new bomb in person at the end of the session on 24 July. He walked across the room and casually informed Stalin that the United States now had a new weapon. Stalin knew all about the development of the atomic bomb through his spy network. He replied equally casually: 'Good, I hope the United States will use it.' Truman thought Stalin had not understood. But Uncle Joe realised immediately the significance of what he had been told. That evening, he instructed Molotov to speed up the development of the Soviet bomb.

The following day, Churchill flew back to London and the conference was put on hold while he awaited the results of the general election. Some party estimates suggested that the Conservatives would win with a majority of between fifty and eighty seats. On the morning of 26 July, Churchill followed the news of the results in his Map Room. One after another safe Conservative seat fell to Labour. By midday, it was becoming clear that the Labour Party had won a landslide victory. The decision of the British people was absolutely clear. Churchill was out.

Clementine was secretly relieved. After more than five years of relentless pressure, she wanted Winston to have a break rather than face the overwhelming challenges of managing the peace. Churchill, on the other hand, was utterly devastated by the news of his defeat. He could not understand how the very people he had led to victory, and who had cheered and mobbed him a few weeks before, had now rejected him. He was seventy years old. This might be the end of his political career. At lunch, Clementine told her husband that the result 'may well be a blessing in disguise'. Churchill replied: 'At the moment it seems quite effectively disguised.'[26] Churchill resigned that evening and Attlee, with a majority of 146, formed a Labour government. It would be one of the truly great reforming governments in British history.

On 6 August, an atom bomb was dropped on the Japanese city of Hiroshima. It exploded with the force of 13,000 tons of TNT. The heat it generated was so intense that it melted bricks and roof tiles, and incinerated human beings so completely that nothing remained of them except light outlines on scorched pavements. About a hundred thousand civilians died within hours. Thousands more died of radiation poisoning over the next few months and years. On 8 August, Stalin declared war on Japan and the Red Army entered Manchuria and then Korea, at that time a part of Japan. On 9 August, with minimal strategic need, a second atom bomb was dropped on the city of Nagasaki. The following day, the Japanese Emperor announced his intention to surrender. Terms were agreed shortly after and the Japanese surrender was formally signed a few weeks later on board the battleship USS *Missouri* in Tokyo harbour. The Second World War was over. The atomic age had begun. On that same day, Churchill at last took off on a holiday.

11

Churchill's War

Winston Churchill remains one of the most fascinating figures of the twentieth century. Aware of how history judges individuals, he was determined to be one of the first to write his own account of the war years. He said he would be happy for the judgement of events to be left to history – but that he would be one of the historians.[1] The massive six-volume memoir–history of the Second World War that he published from 1948 to 1954 presented his own interpretation of events very clearly and lucidly. Backing up his narrative, he provided a mass of documents he had special clearance from the Cabinet to publish.[2] Up to his death, most of the memoirs and accounts that were published painted a positive picture of his leadership, with the exception of Arthur Bryant's working of General Brooke's diaries. These first appeared in Bryant's two books *The Turn of the Tide 1939–43* and *Triumph in the West 1943–46* in the late 1950s. Churchill was extremely hurt and offended by the publication of these diary entries frequently written in anger and despair while events were still unfolding, often late at night, and without the mediation and reflection of time. According to John Colville, Churchill deliberately and ostentatiously turned his back on Brooke, by then Viscount Alanbrooke, after the publication of Bryant's books. Colville said he knew of no other person who received similar treatment from Churchill.[3] But after Churchill's death, the pendulum began to swing and several historians took a more robust and often hostile view of the great man.[4] More recently, the pendulum has swung back again. In 2002, Churchill was voted 'Greatest Briton' in a BBC television series.

One of the key questions when it comes to evaluating his leadership of Britain during the Second World War is: what difference did he make? Churchill later said that all he did was 'to let the Lion roar', that the British people were the lion-hearted nation and it was his good luck to lead them.[5] But in this, rarely for once, he underplays his own role. In May and June 1940, when France was disastrously overrun in a matter of weeks, when Britain faced the possibility of invasion and attack from the air, and when the sea-lanes that kept the nation functioning might have been cut, the situation seemed truly hopeless. At that time, what was the real view of the British people? We will never know for sure, but Churchill had only a fragile hold on Downing Street. Most members of the Conservative Party, by far the largest number of MPs in the Commons, were far from convinced that he was the right man to lead the nation in its hour of peril. Moreover, there is plenty of evidence to suggest that if another leader, perhaps Lord Halifax, had told the British people at that moment that the only sensible option was to make peace with Germany, then Britons might well have eagerly gone along with him. And as several commentators have said, the ensuing history of Europe would have taken a very different course. As it was, Churchill, with his strong sense of history and an equally strong sense of destiny, would have 'no parley' with the Nazis. His speeches and his courage at this moment without question had an immense impact on the course of events. First to the War Cabinet, then to the full Cabinet, then to Parliament, and then to the people at large, Churchill provided a direction and a leadership that filled them with pride and the resolution to fight on against an evil regime. This was not just a question of clever oratory, of fine eloquence, it was a question of saying what needed to be said. Hitler had to be stopped, and the only way this was possible was by defeating him in war. Had Churchill dropped dead of a heart attack a few months later, as he could well have done, his contribution to world history would still have been immense. But as it was he went on saying what needed to be said to keep the British people determined to fight on against what looked like impossible odds, until first the Soviet Union and then the United States joined the struggle. From May 1940 until November 1942, he kept the war effort going before the contribution of those new allies really made a difference to the balance of the war. He stayed put in London when it was wise to do so. He visited bomb-damaged cities when he needed to. And he visited the Allies and the military front in a seemingly endless round of travels in order to understand better what was going on and to inspire and motivate those he met.

In the latter part of the war, as Britain's contribution to the defeat of fascism in Europe and militarism in Japan was far exceeded by those of the United States and the Soviet Union, inevitably Churchill's ability to make a difference lessened. But through his Chiefs of Staff he convinced the Americans to postpone D-Day until 1944, a postponement which few historians today regard as anything other than essential for final victory. And Churchill was the first leader to identify the Soviet Union as the next antagonist of the West although he failed to convince Roosevelt of this. But in the years to come Americans would see him almost as a prophet for this vision.

As a military man, he certainly had courage, and luck. He understood how armies and navies worked and he was able to get things done. He had closely observed and reflected on the process of government during war, and when he came to lead he knew exactly how he wanted to structure things. He knew the strengths and weaknesses of experts and advisers. He did not want to be given an over-rosy interpretation of events; nor did he want to hide disasters from the public when they happened, as they did in plenty. But even when conveying bad news he had the ability to exude confidence.[6] The biggest criticism against him on this point is that he listened too unquestioningly to Lord Cherwell, his principal scientific aide. Cherwell was to be proved wrong on many key matters – from the effectiveness of bombing as a way to destroy the German war economy to the ability of the Germans to produce rocket missiles. Even though he was over-reliant on this single individual at the expense of other great wartime scientists, Churchill valued science and technology highly, at least as far as they could improve military actions, at a time when there was much suspicion of scientists. And he liked fresh, unconventional, even unorthodox thinking. He could inspire people. His words inspired a generation living through the war. No matter how old-fashioned or out of place he seemed in 1940, he said what people wanted to hear in those extreme circumstances. He made people feel important and that their lives were linked to a long and great history. He had the ability to push individuals, sometimes beyond what they thought they were capable of.

This account has been full of quotes from men and women who felt lit up in his presence. R.V. Jones, the young scientist who first met the Prime Minister when he arrived late for the crucial Downing Street meeting that ushered in the Battle of the Beams in June 1940, had several meetings with Churchill through the war. Thirty years later, reflecting on being with Churchill, he wrote: 'I had the feeling of being recharged by contact with a source of living power. Here was strength, resolution, humour, readiness to listen, to ask the searching question and when convinced, to act.'[7] Even accounting for the impression an elder statesman would naturally make on a younger man, this describes a rare quality in leadership. Churchill had led a team of scientists and military chiefs, he had cut through red tape, encouraged innovation and fresh thinking. The War Lab he developed around him helped Britain survive and did much to contribute to victory.

The other side of this is that he cajoled and sometimes bullied his military chiefs beyond what was fair and reasonable, although not usually beyond endurance. He usually knew when to stop, and when a kind word was needed. When pushing one of his secretaries, Elizabeth Layton, at a critical moment late at night he paused, realising her exhaustion, and said: 'We must go on like the gun horses until we drop.'[8] Of course she was then willing to go that extra mile for him and with him. He probably came nearest to a full-scale falling out with General Brooke, particularly when exhaustion had reduced both men to a fragile state in the spring of 1944, and they argued intensely about future strategy. But even here, Churchill did not push Brooke into resignation, although he and the Chiefs of Staff came near to it. And, as Brooke's diary recorded, Churchill was keen to make up and express his appreciation for his CIGS when he felt he had gone too far.

As we have seen throughout this book, Churchill's intervention was critical at many key moments. From getting extra resources for the much-needed expansion of the code-breaking centre at Bletchley Park to focusing both government and military thinking on solving the U-boat threat; from giving the go-ahead to constructing the floating Mulberry harbours to pausing the bombing offensive until navigation techniques could be improved. These are just a few of his many key interventions. And of course, more than anything else, Churchill realised the importance of the 'special relationship' with the United States. This awareness began well before he came to office but it became a life-saver during the Spitfire Summer of 1940. Despite his differences with President Roosevelt, differences that have often been overlooked in the rosy glow of victory, Churchill worked immensely hard in building not just a strong personal relationship but a system of collaboration between two military operations that not only survived the war but became a key alliance of the post-war world.

When it came to shaping military strategy, Churchill's record is more chequered. He has been criticised a lot for his 'dispersionist' strategy, always looking for a flank to attack, for his obsession with dispersing his forces to surprise the enemy by attacking where they least expect it. This, in essence, was what the Dardanelles campaign was all about. And, in part, it was what was behind his Mediterranean strategy. But sometimes it worked brilliantly. In 1941, there was nowhere else realistically to attack Axis forces than in North Africa. By dragging Hitler into this theatre of war, and later into Italy to defend his Italian ally, and by diverting his forces from elsewhere, Churchill's strategy worked. He had a canny ability to sense where the enemy was weak and to attack there. But Nigel Knight, among others, believes that, as an overall strategy, this is fundamentally flawed and wars are won by concentrating forces against the enemy, not by dispersing them. Knight claims that Churchill might have extended the war by as much as a year by pursuing this strategy.[9] Of course, this can never be proven. Churchill's unrealised pet project, to attack Norway, probably helped to keep a dozen German divisions in that country when they could have turned the tide of battle elsewhere. On the other hand, his plan to invade Sumatra in 1944 seems ridiculous today.

There is also the controversy of the bombing offensive and accusations have even been made that by ordering the area bombing of German towns and cities Churchill was effectively a war criminal. But it is profoundly *un*-historical to apply the moral criteria of one age to individuals living in a different era. Today, when war deaths are reported in the mass media and in Parliament name by name, it is difficult to imagine a context in which thousands or even tens of thousands of people lost their lives in a single day. So it would be historically wrong to condemn Churchill for being the political leader of a nation that left a trail of such devastating loss of life across German cities. At the time, in the aftermath of Hitler's Blitz on British cities, there was no moral condemnation of the bombing of Germany, just a debate as to whether the ends (the partial destruction of the German war economy) justified the means (the vast allocation of resources within the US and British war economies and the huge loss of life among the bomber crews). Later, this became a political and a moral issue. Political in that no campaign medal was ever given to the men of Bomber Command. Moral in that it has been said that the killing of civilians in war is *always* wrong. But it still goes on today despite our moral revulsion. And the moral debate about Churchill continues.

Did Churchill have doubts? Certainly he did. He would not have been human otherwise. On the day he became Prime Minister, driving back to the Admiralty, his personal detective offered his congratulations and Churchill responded grimly

that he hoped it was not too late. Returning from his penulti-
mate meeting with the French leaders in June 1940, he turned
to General Ismay and said: 'We fight alone.' When Ismay tried
to cheer him up by saying he was glad of it, and that 'We'll win
the Battle of Britain,' Churchill turned to him, gave him a look
and said: 'You and I will be dead in three months' time.'[10] On
many occasions over the following years he was brought low
by depression, his 'black dog'. In the First World War, he had
admitted to 'terrible and reasonless depressions'.[11] In the
Second, he was worried by recurring fears that, for instance, the
British soldier no longer had the warrior spirit to fight with the
determination that would bring victory. He was also terribly
upset on hearing news of losses, particularly at sea. Despite the
strength of his oratory and the courage he displayed, he wore
his emotions on his sleeve for most of the war. There are dozens
of accounts of tears falling down his cheeks at emotional
moments, not something it is easy to imagine with more recent
leaders.

The telephone calls Churchill frequently had with Roosevelt
over the first ever 'hot-line', which was set up between the
Cabinet War Rooms and the Oval Office, had to be listened to
by a censor because they were sent by radio and although
scrambled they could be intercepted by the Germans (as indeed
they were). Ruth Ive, then in her early twenties, was one
of those who listened in to these phone calls and she was
instructed to interrupt the two leaders if certain security pro-
tocols were breached. She could always tell when Churchill was
in a depressed frame of mind. She was always alarmed by hearing him in this depressed state, wondering at times if he were
capable of carrying on.[12] Many others who worked with him
closely had the same worries.

So Churchill was only human after all, which should be no
surprise. He needed the right people around him to bring out

the best in him, and for him to bring out the best in them. And they in turn passed on his passion and his inspiration to others. That was what his War Lab was all about. So, yes, he really did make a difference. At a time of a national, European and world crisis, when leadership mattered, Britain for once had the right man in the right place doing the right job.

Notes

Introduction

1. This account is based on Winston Churchill *My Early Life* pp. 12–20.
2. This account is based on Winston Churchill *My Early Life* pp. 179–93.
3. This account is taken from John Wheeler-Bennett *King George VI* pp. 601–5 and from Winston Churchill *Second World War Vol. V* pp. 546–51. The King's letter to Churchill is quoted in both.
4. Ronald Clark *The Rise of the Boffins* p. 162, and see p. 177–8 of this book.
5. R.V. Jones 'Churchill and Science' in Robert Blake and Wm. Roger Louis (eds) *Churchill* p. 437.
6. Piers Brendon *Winston Churchill: A Brief Life* pp. 148–50.
7. Winston Churchill *The Second World War Vol. I* pp. 526–7.

Chapter 1 – Preparation: The Army and the Navy

1. See David Cannadine *The Decline and Fall of the British Aristocracy* pp. 113 and 397–8.
2. Churchill always claimed he was born premature. He could have been conceived before his parents were married in April 1874. We will never know which is true.
3. Winston Churchill *My Early Life* pp. 4–5.
4. Winston Churchill *My Early Life* pp. 15–16.
5. Winston Churchill *My Early Life* p. 76.
6. Randolph Churchill *Winston S. Churchill Vol I. Youth: 1874–1900 Companion Volume II 1896–1900* p. 930.

7. Randolph Churchill *Winston S. Churchill Vol. I: Youth 1874–1900* p. 418.

8. This book does not attempt to be any sort of political biography of Churchill, nor to provide a critique of his political views. Of recent biographies, the best political account comes in Roy Jenkins *Churchill*.

9. Randolph Churchill *Winston S. Churchill Vol. I: Youth* p. 463.

10. Winston Churchill *My Early Life* p. 248.

11. Winston Churchill *My Early Life* p. 256.

12. Winston Churchill *My Early Life* p. 277.

13. The whole story of his capture and escape is told in detail in Celia Sandys *Churchill Wanted Dead or Alive*. Sandys, his granddaughter, corrects some of the myths Churchill himself created about this episode.

14. Randolph Churchill *Winston S. Churchill Vol. I: Youth 1874–1900* p. 524.

15. Martin Gilbert *Churchill: A Life* p. 85.

16. Winston Churchill *My Early Life* p. 298.

17. Randolph Churchill *Winston S. Churchill Vol. II: Young Statesman* p. 71.

18. Randolph Churchill *Winston S. Churchill Vol. II: Young Statesman* p. 228.

19. Clementine had a Higher School Certificate (roughly equivalent to A Levels today) in French, German and Biology from Berkhamsted High School. Her headmistress wanted her to go on to university but her mother decided when she was eighteen that she had had enough education and that it was time to enter society and find a husband. She had had two engagements, both broken off by her, before she met Churchill. See Mary Soames *Clementine Churchill* pp. 22–34.

20. A selection of these letters, which beautifully captures the ups and downs of their relationship, features in *Speaking for Themselves: The Personal Letters of Winston and Clementine Churchill*, edited by their daughter Mary Soames.

21. Winston Churchill *The World Crisis Vol. I* p. 49 (this and all subsequent volume and page references are from the 1939 edition of *The World Crisis*).

22. A variation of this appears in Winston Churchill *The World Crisis Vol. I* p. 24: 'The Admiralty had demanded six ships; the economists offered four; and we finally compromised on eight.'

23. Winston Churchill *The World Crisis Vol. I* pp. 51–70, quoted in Stephen Roskill *Churchill and the Admirals* p. 30.
24. Randolph Churchill *Winston S. Churchill Vol. II: Young Statesman* p. 703.
25. Randolph Churchill *Winston S. Churchill Vol. II: Young Statesman* p. 686.
26. Winston Churchill *The World Crisis Vol. I* pp. 155–7.
27. Randolph Churchill *Winston S. Churchill Vol. II: Young Statesman* p. 710.
28. Martin Gilbert *Winston S. Churchill Vol. III* p. 31.

Chapter 2 – Preparation: The War and the Wilderness

1. David Kahn *Seizing the Enigma* pp. 15ff. and David Stafford *Churchill and Secret Service* pp. 70ff.
2. Winston Churchill *World Crisis Vol. II* p. 562.
3. Winston Churchill *World Crisis Vol. I* p. 414.
4. Nicholas Rankin *Churchill's Wizards* pp. 13–14.
5. Letter from the Master General of the Ordnance, 26 February 1915, quoted in Winston Churchill *The World Crisis Vol. II* p.512.
6. Martin Gilbert *Winston S. Churchill Vol. III: 1914–16* p. 537; see also Winston Churchill *The World Crisis Vol. II* pp. 508–16.
7. Winston Churchill *The World Crisis Vol. II* p. 466.
8. Martin Gilbert *Winston S. Churchill Vol. III: 1914–16* p. 465.
9. Martin Gilbert *Winston S. Churchill Vol. III: 1914–16* pp. 579–80 and 574.
10. Martin Gilbert *Winston S. Churchill Vol. III: 1914–16* pp. 610, 686, 705 and 745.
11. Martin Gilbert *Winston S. Churchill Vol. III: 1914–16* pp. 609 and 748.
12. Gary Sheffield and John Bourne (eds) *Douglas Haig: War Diaries and Letters* pp. 315 and 371.
13. Winston Churchill *The World Crisis Vol. IV* p. 543.
14. The book is part history and part autobiography of Churchill's own contribution to events from the time he was appointed First Lord of the Admiralty in 1911. His political opponent Bonar Law cruelly but amusingly described the book as 'an autobiography disguised as an history of the universe'.
15. This line was remembered by Lady Violet Bonham Carter; see Martin Gilbert *Winston S. Churchill Vol. IV: 1916–1922* p. 278.

16. Martin Gilbert *Winston S. Churchill Vol. IV: 1916–1922* p. 332.

17. Winston Churchill *Thoughts and Adventures* p. 213.

18. Paul Addison *Churchill on the Home Front*, p. 243.

19. Piers Brendon *Winston Churchill: A Brief Life* p. 100.

20. Chartwell is today owned by the National Trust and is open to visitors; see http://www.nationaltrust.org.uk/main/w-chartwell.htm.

21. Piers Brendon *Winston Churchill: A Brief Life* p. 118.

22. Maurice Ashley *Churchill as Historian* p. 18.

23. See J.H. Plumb 'The Historian' in A.J.P. Taylor et al. *Churchill: Four Faces and the Man* p. 119–29. Plumb rightly draws a distinction between Churchill's histories of events with which he was not personally involved and his histories of the two world wars, which, as has already been noted, are much more autobiographical and so have great value as guides to Churchill's thinking at the times they describe.

24. Geoffrey Best *Churchill: A Study in Greatness* pp. 146–7 and David Stafford *Churchill and Secret Service* p. 180.

25. Hansard *Parliamentary Debates* House of Commons, 21 February 1938.

26. Martin Gilbert *Winston S. Churchill Vol. V: 1922–39* p. 917.

27. Hansard *Parliamentary Debates* House of Commons, 5 October 1938.

28. Martin Gilbert *Winston S. Churchill Vol. V: 1922–39* p. 1106.

29. For instance, the broadcast features in the Ministry of Information documentary *The First Days* produced by Alberto Cavalcanti, directed by Humphrey Jennings, Harry Watt, Pat Jackson and others.

30. Martin Gilbert *Winston S. Churchill Vol. VI: Finest Hour* p. 4.

31. Winston Churchill *The Second World War Vol. I* p. 320. Some scholars have suggested that this could have been a message of warning rather than of welcome, for example Geoffrey Best *Churchill and War* p. 107. Other scholars have doubted that the message was ever sent. Whether apocryphal or not, it is entirely believable.

32. Sir Alan Brooke quoted in Nicholas Rankin *Churchill's Wizards* p. 237.

33. Martin Gilbert *Winston S. Churchill Vol. VI: Finest Hour* pp. 156–7.

34. Martin Gilbert *Winston S. Churchill Vol. VI: Finest Hour* p. 62. The story of the sinking of HMS *Royal Oak* and the dreadful loss of life soon became known. However, the Admiralty did not reveal the number of boy sailors who went down with the ship. It did, though, change the policy with regard to boy sailors, who were henceforth no longer allowed to serve on ships that were on active duty.

35. Peter Calvocoressi and Guy Wint *Total War* p. 100.

36. John Keegan (ed.) *Churchill's Generals* p. 27.

37. For instance, from Bob Boothby and from Harold Macmillan; see Martin Gilbert *Winston S. Churchill Vol. VI: Finest Hour* pp. 302ff. and Harold Macmillan *The Blast of War 1939–45* p. 74.

38. From the diary of Henry 'Chips' Channon, quoted in Martin Gilbert *Winston S. Churchill Vol. VI: Finest Hour* p. 294.

39. Hansard *Parliamentary Debates* House of Commons, 8 May 1940.

40. Winston Churchill *The Second World War Vol. I* pp. 523–4.

41. From John Colville's interview for *The World at War* Thames Television documentary series, 1973, producer Jeremy Isaacs. See also John Colville *The Fringes of Power – Downing Street Diaries 1939–45* p. 123.

42. Churchill wrongly dates the meeting as being on 10 May, whereas all other accounts place it on the 9th. Also, Churchill forgets that David Margesson, the Tory Chief Whip, was present. Halifax's diary is quoted in Robert Blake 'How Churchill Became Prime Minister' in Robert Blake and Wm. Roger Louis (eds) *Churchill* pp. 265–7. Other accounts have led Andrew Roberts to give another version of these critical few hours in *Hitler and Churchill: Secrets of Leadership* pp. 94–100. Roberts, who is also Halifax's biographer, recounts several discussions between Chamberlain and Halifax, in which the Prime Minister proposed Halifax as his successor and in which Halifax literally felt pain in the pit of his stomach. Roberts also quotes the recently published diaries of the American ambassador, Joseph Kennedy, who discussed the meeting with Chamberlain later. In both Blake's and Roberts's accounts, Chamberlain is more active in trying to prevent Churchill from becoming Prime Minister, and Churchill is more determined to grasp the position.

43. Martin Gilbert *Winston S. Churchill Vol. VI: Finest Hour* p. 305.

Chapter 3 – Action This Day

1. The King's diary is quoted in John Wheeler-Bennett *King George VI: His Life and Reign* p. 444. Churchill's account is in Winston Churchill *The Second World War Vol. I* p. 525.

2. Winston Churchill *The Second World War Vol. I* pp. 525–7.

3. Richard Broad and Suzie Fleming (eds) *Nella Last's War* p. 55.

4. Winston Churchill *The Second World War Vol. I* p. 527.

5. John Colville in Lord Normanbrook (and others) *Action This Day: Working with Churchill* p. 48.

6. Hansard *Parliamentary Debates* House of Commons, 13 May 1940.

7. Hastings Ismay *Memoirs* p. 158.

8. John Colville *The Churchillians* pp. 205–6.

9. John Keegan (ed.) *Churchill's Generals* p. 27.

10. Hastings Ismay *Memoirs* p. 159.

11. John Colville *The Churchillians* p. 146.

12. Ronald Lewin *Churchill as Warlord* pp. 41–2.

13. Lord Normanbrook *Action This Day* p. 22.

14. Elizabeth Nel [née Layton] *Mr Churchill's Secretary* pp. 29–30.

15. The Cabinet War Rooms, restored as they were left at the end of the war, can be visited today as a part of the Imperial War Museum. See: http://cwr.iwm.org.uk/server/show/nav.221

16. John Keegan (ed.) *Churchill's Generals* p. 7.

17. Many of those who worked closely with Churchill describe the tough, demanding hours he worked, for instance John Colville *The Fringes of Power passim* and *The Churchillians* pp. 64ff.; Field Marshal Lord Alanbrooke *War Diaries passim*; and Elizabeth Nel (née Layton) *Mr Churchill's Secretary* pp. 27ff.

18. Hastings Ismay *Memoirs* pp. 164–5 and John Colville *The Churchillians* p. 146.

19. Martin Gilbert *Winston S. Churchill Vol. VI: Finest Hour* pp. 339–40 and Winston Churchill *The Second World War Vol. II* p. 38–9

20. Winston Churchill *The Second World War Vol. II* p. 42.

21. There has been much debate about Hitler's intervention confirming von Rundstedt's halt order; see, for instance, Andrew Roberts *Hitler and Churchill: Secrets of Leadership* pp. 105–7 and *The Storm of War* pp. 60–4. Roberts does not accept the argument that Hitler allowed the British Expeditionary Forces to escape in order to get better terms in negotiations with Britain.

22. John Lukacs *Five Days in London May 1940* p. 19.

23. Minutes of the War Cabinet, ref: CAB 65/13, WM 145.

24. In *The Second World War* Churchill does not relate the events of the Halifax challenge to his leadership, perhaps not wanting to cast a shadow over Halifax's reputation, perhaps because he felt awkward about having been nearly deflected at this critical point. He does recount the support he received at the full Cabinet meeting on the

afternoon of 28 May 1940 in *Vol. II* pp. 87–8. The account of the meeting in Hugh Dalton's diaries and his memoirs is quoted in John Lukacs *Five Days in London May 1940* pp. 4–5 and 183–4. And in the same book Lukacs talks of the momentousness of this day on p. 2.

25. Hansard *Parliamentary Debates* House of Commons, 4 June 1940.

26. Martin Gilbert *Winston S. Churchill Vol. VI: Finest Hour* p. 469. It has been claimed by David Irving in *Churchill's War Vol. I* p. 313 that the BBC transmitted a version of this speech that evening that was not read by Churchill but by an actor, Norman Shelley, well known at the time for playing Larry the Lamb in the BBC's *Children's Hour*. In fact, extracts of the speech on the BBC Home Service News that evening were read by the newsreader, as the Vita Sackville-West comment makes clear. It is possible that Shelley recorded a version of the speech for transmission overseas, but this was not heard in the UK. See D.J. Wenden 'Churchill, Radio, and Cinema' in Robert Blake and Wm. Roger Louis (eds) *Churchill* pp. 236–7.

27. Winston Churchill *The Second World War Vol. II* pp. 136–42 relates the whole story of this trip to France and Churchill's return.

Chapter 4 – Spitfire Summer

1. Hansard *Parliamentary Debates* House of Commons, 18 June 1940.

2. John Colville *The Fringes of Power* p. 165. Listening to the speech held in the BBC Archives today, he does not come across as tired, nor sound like he is smoking a cigar. But by this point Colville was working with Churchill virtually all day, every day, so his impression is interesting. One of Colville's dinner companions, who listened to the speech on the radio with him, thought that Churchill sounded like 'a bishop'!

3. Martin Gilbert *Winston S. Churchill Vol. VI: Finest Hour* p. 571.

4. D.J. Wenden 'Churchill, Radio, and Cinema' in Robert Blake and Wm. Roger Louis (eds) *Churchill* p. 238. They were recorded by Churchill in 1949 and were released by Decca as a set of LPs. Some of the versions of Churchill's wartime speeches that are available today are these later recordings.

5. Warren Kimball *Forged in War* p. 15 and David Stafford *Roosevelt and Churchill* p. xvi.

6. Martin Gilbert *Winston S. Churchill Vol. VI: Finest Hour* p. 52.

7. Warren Kimball *Forged in War* pp. 3ff.

8. Martin Gilbert *Winston S. Churchill Vol. VI: Finest Hour* p. 146.

9. Martin Gilbert *Winston S. Churchill Vol. VI: Finest Hour* pp. 345–6.

10. Ronald Lewin *Churchill as Warlord* p. 37.

11. Martin Gilbert *Winston S. Churchill Vol. VI: Finest Hour* p. 356.

12. Martin Gilbert *Winston S. Churchill Vol. VI: Finest Hour* pp. 427 and 689.

13. Winston Churchill *The Second World War Vol. II* p. 206.

14. Ivan Maisky *Memoirs of a Soviet Ambassador: The War 1939–43* pp. 99–100.

15. John Colville *The Fringes of Power* p. 200.

16. For the Spitfire story, see Taylor Downing and Andrew Johnston *Battle Stations* pp. 11–35 and Leo McKinstry *Spitfire: Portrait of a Legend*.

17. Quote from Bob Doe, the third-highest-scoring ace in the Battle of Britain, in *Battle Stations – Spitfire Squadron*, producer Taylor Downing, director Andrew Johnston, Flashback Television, 2000; and quoted in Taylor Downing and Andrew Johnston *Battle Stations* p. 44. See also Bob Doe *Fighter Pilot*.

18. Constance Babington Smith *Evidence in Camera* p. 62.

19. The evening is recounted in James Marshall-Cornwall *Rumours of War* pp. 166–71. See also Martin Gilbert *Winston S. Churchill Vol. VI: Finest Hour* pp. 682–5.

20. Hastings Ismay *Memoirs* pp. 179–80.

21. The speech is from Hansard *Parliamentary Debates* House of Commons, 20 August 1940; see John Colville *The Fringes of Power* p. 227 for the anecdote about the car journey.

22. Violet Bonham Carter in Martin Gilbert *Winston S. Churchill Vol. VI* p. 742.

23. Hastings Ismay *Memoirs* pp. 183–4.

24. Winston Churchill *The Second World War Vol. II* pp. 293–7.

25. For instance, in Angus Calder *The People's War: Britain 1939–45* and in the spate of books it ushered in. And more recently in Juliet Gardiner *Wartime Britain 1939–45* pp. 530ff.

26. David Cannadine (ed.) *Winston Churchill: Blood, Toil, Tears and Sweat; The Great Speeches* p. xiv.

27. Quoted in David Cannadine (ed.) *Winston Churchill: Blood, Toil, Tears and Sweat; The Great Speeches* p. xxxiii.

28. Asa Briggs *The War of Words* pp. 187 and 297.

29. From a speech given at Westminster Hall, London, 30 November 1954 on the occasion of a parliamentary tribute to his eightieth birthday – he was the first prime minister since Gladstone to be in power at the age of eighty.

30. Asa Briggs *The War of Words* p. 10.
31. E. Bliss (ed.) *In Search of Light: The Broadcasts of Edward R. Murrow* p. 237.
32. David Reynolds '1940: The Worst and Finest Hour' in Robert Blake and Wm. Roger Louis (eds) *Churchill* p. 254.
33. John Colville *The Fringes of Power* p. 217.

Chapter 5 – The Wizard War

1. Sir Solly (later Lord) Zuckerman *Scientists and War* p. 9.
2. See Winston Churchill *The World Crisis Vol. II* pp. 508–26.
3. Winston Churchill *The World Crisis Vol. II* p. 464.
4. Thomas Wilson *Churchill and the Prof* p. 12.
5. Winston Churchill *The Second World War Vol. II* p. 338.
6. Lord Birkenhead *The Prof in Two Worlds* p. 159.
7. Thomas Wilson *Churchill and the Prof* pp. 29–30.
8. Hansard *Parliamentary Debates* House of Commons, 10 November 1932.
9. Thomas Wilson *Churchill and the Prof* p. 33.
10. Ronald Clark *The Rise of the Boffins* p. 161; see also Ronald Clark *Tizard passim*.
11. Ronald Clark *The Rise of the Boffins* p. 51; see also Ronald Clark *Tizard* pp. 149–63.
12. David Kahn *Seizing the Enigma* pp. 68ff.
13. See F.H. Hinsley and Alan Stripp (eds) *Code Breakers: The Inside Story of Bletchley Park passim* and Michael Smith *Station X* pp. 16ff.
14. Ronald Lewin *Ultra Goes to War* p. 183.
15. Martin Gilbert *Winston S. Churchill Vol. VI: Finest Hour* pp. 611–13, and Ronald Lewin *Churchill as Warlord* p. 75 and *Ultra Goes to War* p. 64.
16. Michael Smith *Station X* p. 78.
17. Martin Gilbert *Winston S. Churchill Vol. VI: Finest Hour* pp. 1185–6 and Michael Smith *Station X* pp. 79–81.
18. Winston Churchill *The Second World War Vol. II* p. 340. The meeting is also recounted in R.V. Jones *Most Secret War* pp. 100–5.
19. R.V. Jones *Most Secret War* pp. 101–2 and Martin Gilbert *Winston S. Churchill Vol. VI: Finest Hour* pp. 380–2.
20. R.V. Jones *Most Secret War* pp. 149–51, and John Colville *The Fringes of Power* pp. 294–5 and *The Churchillians* pp. 62–3.
21. Winston Churchill *The Second World War Vol. II* p. 343. There are several accounts of the Battle of the Beams, including Ronald Clark

The Rise of the Boffins pp. 98–125, Brian Johnson *The Secret War* pp. 11–61 and R.V. Jones *Most Secret War* pp. 92–188.

22. Ronald Clark *The Rise of the Boffins* p. 136; see also Ronald Clark *Tizard* pp. 248-252.

23. The official historian was James Phinney Baxter, quoted in Robert Buderi *The Invention that Changed the World* pp. 27 and 36–7 and Ronald Clark *Tizard* p. 268.

24. Ronald Clark *The Rise of the Boffins* p. 215.

25. *The Origins and Development of Operational Research in the Royal Air Force* HMSO Air Ministry Publication 3368, 1962.

26. For instance, the author's father, Peter Downing, was recruited into the RAF on graduation from King's College, London, with a mathematics degree, made a Pilot Officer, given radar operator's wings and sent immediately by Imperial Airways flying boat to Cairo, a sign of the urgent need for Operational Research mathematicians in the Middle East.

27. See Sir (later Lord) Solly Zuckerman's memoirs, *From Apes to Warlords*. Zuckerman later became Chief Scientific Adviser to the Ministry of Defence (1960–4) and then Chief Scientific Adviser to the British Government (1964–71). Despite his senior position he was an outspoken critic of the policy of nuclear deterrence.

28. Ronald Clark *The Rise of the Boffins* p. 162.

29. Tom Shachtman *Laboratory Warriors* p. 157.

30. Jeremy Isaacs and Taylor Downing *Cold War* p. 18.

Chapter 6 – The Generals

1. Ronald Lewin *Churchill as Warlord* p. 84.

2. Martin Gilbert *Winston S. Churchill Vol. VI: Finest Hour* p. 667.

3. John Keegan *Churchill* pp. 127–8 and 'Churchill's Strategy' in Robert Blake and Wm. Roger Louis (eds) *Churchill* pp. 333–4.

4. Patrick Delaforce *Churchill's Secret Weapons* p. 45 and Martin Gilbert *Winston S. Churchill Vol. VI: Finest Hour* pp. 746–7.

5. Winston Churchill *The Second World War Vol. II* p. 379.

6. Martin Gilbert *Winston S. Churchill Vol. VI: Finest Hour* pp. 981–1000.

7. David Stafford *Churchill and Secret Service* pp. 231–3.

8. Robert E. Sherwood *The White House Papers of Harry L. Hopkins Vol. I* pp. 256–7.

9. Winston Churchill *The Second World War Vol. III* p. 111, and Martin Gilbert *Winston S. Churchill Vol. VI: Finest Hour* pp. 1031–3 and *Churchill: A Life* p. 254.
10. Ronald Lewin *Churchill as Warlord* p. 57.
11. David Stafford *Churchill and Secret Service* pp. 250–4.
12. Ronald Lewin *Churchill as Warlord* p. 73 and Winston Churchill *The Second World War Vol. III* p. 223.
13. John Keegan (ed.) *Churchill's Generals* p. 80 and Hastings Ismay *Memoirs* pp. 269–71.
14. John Colville *Fringes of Power* p. 404. Colville and Churchill also discussed Wavell's dismissal in this after-dinner conversation in the garden. Colville told Churchill that Wavell would probably write his memoirs after the war and put his own side of the story of his dispute with the Prime Minister. Churchill replied that he would write his version as well 'and would bet he sold more copies'! This amusing aside proves that Churchill was thinking about writing his own account of the war as early as 1941.
15. Winston Churchill *The Second World War Vol. III* pp. 331–3.
16. Winston Churchill *The Second World War Vol. III* pp. 320–3. In this account, written in the late 1940s, Churchill was careful not to reveal the Ultra secret. He used phrases like 'Intelligence reports from one of our most trusted sources' and spoke of 'reliable agents' in the neutral countries. The story of breaking the Enigma codes was not made public until the 1970s. See also David Stafford *Churchill and Secret Service* pp. 258–9.
17. Sir John Martin *Downing Street: The War Years* p. 58.
18. David Fraser *Alanbrooke* p. 202.
19. Alex Danchev and Daniel Todman (eds) *Field Marshal Lord Alanbrooke's War Diaries* pp. 160–1.
20. John Keegan (ed.) *Churchill's Generals* p. 90.
21. Alex Danchev and Daniel Todman (eds) *Field Marshal Lord Alanbrooke's War Diaries* p. 590. In their 'Introduction' to the *War Diaries*, Danchev and Todman explore the circumstances of the writing of the diaries: pp. xiff. The original diaries are held at the Liddell Hart Centre for Military Archives at King's College, London.
22. Martin Gilbert *Winston S. Churchill Vol. VI: Finest Hour* p. 1251.
23. David Stafford *Churchill and Secret Service* p. 261.

368 Churchill's War Lab

24. Eyewitness interview with Colonel Manteuffel in Thames TV's *The World at War* episode 'Barbarossa', producer Jeremy Isaacs, director Peter Batty. See also Richard Holmes *The World at War* p. 191.
25. Winston Churchill *The Second World War Vol. III* p. 538.
26. Winston Churchill *The Second World War Vol. III* pp. 539–40.
27. These three papers are reproduced in Winston Churchill *The Second World War Vol. III* pp. 572–86.
28. Winston Churchill *The Second World War Vol. III* p. 588.
29. The speech is quoted in full in David Cannadine (ed.) *Winston Churchill: Blood, Toil, Tears and Sweat; The Great Speeches* pp. 226–33.
30. Lord Moran *Churchill at War 1940–45* pp. 17–18.
31. Lord Moran *Churchill at War 1940–45* p. 22.
32. Martin Gilbert *Winston S. Churchill Vol. VII: Road to Victory* p. 78.
33. Martin Gilbert *Winston S. Churchill Vol. VII: Road to Victory* p. 67.
34. Winston Churchill *The Second World War Vol. IV* pp. 343–4. General Brooke comments in his diary on the generosity of the US offer of Sherman tanks, which had already been allocated to a US armoured division: see Alex Danchev and Daniel Todman (eds) *Field Marshal Lord Alanbrooke's War Diaries* p. 269.
35. Lord Moran *Churchill at War 1940–45* pp. 57–8.
36. Alex Danchev and Daniel Todman (eds) *Field Marshal Lord Alanbrooke's War Diaries* p. 293. Churchill had initially offered Auchinleck's role to Brooke himself, who had been sorely tempted by 'the finest command I could hope for'. But he felt he lacked experience in desert warfare and was 'able to render better services to my country' by remaining as CIGS. See *Field Marshal Lord Alanbrooke's War Diaries* pp. 293–7.
37. Martin Gilbert *Winston S. Churchill Vol. VII: Road to Victory* p. 254.

Chapter 7 – The Admirals

1. Winston Churchill *The Second World War Vol. II* p. 529.
2. Winston Churchill *The Second World War Vol. III* p. 101.
3. Winston Churchill *The Second World War Vol. III* p. 106 and Ronald Lewin *Churchill as Warlord* p. 61.
4. Churchill liked to include documents that showed him in a good light in his post-war history and this directive is reproduced in full in Winston Churchill *The Second World War Vol. III* pp. 107–8.
5. Stephen Roskill, the official historian of the naval war, is highly critical

of this instruction, which he calls one of the most extraordinary signals of the war. See Stephen Roskill *Churchill and the Admirals* p. 125.

6. Stephen Roskill *Churchill and the Admirals* pp. 178 and 125.
7. Stephen Roskill *Churchill and the Admirals* pp. 183ff. and Richard Ollard 'Churchill and the Navy' in Robert Blake and Wm. Roger Louis (eds) *Churchill* pp. 392–3.
8. The most obvious example of this was in Alan Brooke's war diaries. These first came out when released to Arthur Bryant for his two books *The Turn of the Tide 1939–43* and *Triumph in the West 1943–46* in 1957 and 1959. By contrast, Andrew Cunningham in *Admiral A.B. Cunningham: A Sailor's Odyssey*, Arthur Harris in *Bomber Offensive* and Montgomery in his memoirs were all quite bland about their disagreements with the Prime Minister.
9. Winston Churchill *The Second World War Vol. III* p. 551.
10. Winston Churchill *The Second World War Vol. IV* pp. 43–4.
11. Winston Churchill *The Second World War Vol. IV* pp. 87–8.
12. Winston Churchill *The Second World War Vol. IV* p. 81.
13. Martin Gilbert *Winston S. Churchill Vol. VII: Road to Victory* pp. 57–9.
14. David Kahn *Seizing the Enigma* pp. 195–213.
15. This incident was transformed in the movie *U-571*, written and directed by Jonathan Mostow, 2000. In one of the worst Hollywood distortions of history, *U-571* shows a group of *American* submariners deceiving a U-boat into surrendering so they can capture the vital Enigma machine and code books. The derring-do of the U-boats, their captains and their crews has generated lots of films, from *The Enemy Below* (1957) to *Das Boot* (1981).
16. Sir John Martin *Downing Street: The War Years* p. 97.
17. Alex Danchev and Daniel Todman (eds) *Field Marshal Lord Alunbrooke's War Diaries* p. 361.
18. Harold Macmillan *War Diaries: Politics and War in the Mediterranean, January 1943 to May 1945* p. 9.
19. Lord Moran *Churchill at War* p. 99.
20. Ronald Lewin *Churchill as Warlord* p. 186.
21. P.M.S. Blackett *Studies of War* p. 238.

Chapter 8 – Bombing

1. Constance Babbington Smith *Evidence in Camera* pp. 71–6 and Max Hastings *Bomber Command* pp. 80–3.

2. Max Hastings *Bomber Command* pp. 82–8.
3. Winston Churchill *The Second World War Vol. II* pp. 405–6.
4. Sir Charles Webster and Noble Frankland *The Strategic Air Offensive against Germany Vol. I* p. 179.
5. Winston Churchill *Second World War Vol. IV* p. 250.
6. Sir Charles Webster and Noble Frankland *The Strategic Air Offensive against Germany Vol. I* p. 182.
7. Sir Charles Webster and Noble Frankland *The Strategic Air Offensive against Germany Vol. I* p. 186 and Winston Churchill *The Second World War Vol. III* p. 748.
8. Max Hastings *Bomber Command* p. 135.
9. Sir Arthur Harris *Bomber Offensive* pp. 151–5 and Henry Probert *Bomber Harris* pp. 133–4.
10. Cherwell's minute is quoted at length in Thomas Wilson *Churchill and the Prof* p. 74.
11. Ronald Lewin *Churchill as Warlord* p. 101.
12. Sir Charles Webster and Noble Frankland *The Strategic Air Offensive against Germany Vol. I* p. 336.
13. Winston Churchill *The Second World War Vol. IV* p. 433.
14. Winston Churchill *The Second World War Vol. IV* p. 443.
15. Winston Churchill *The Second World War Vol. IV* p. 253.
16. Sir Charles Webster and Noble Frankland *The Strategic Air Offensive against Germany Vol. II* pp. 12–13.
17. Martin Gilbert *Winston S. Churchill Vol. VII: Road to Victory* pp. 295 and 302.
18. Martin Gilbert *Winston S. Churchill Vol. VII: Road to Victory* p. 303.
19. Max Hastings *Bomber Command* p. 189.
20. This film is held at the Landesmedienzentrum in Hamburg. Clips of the film are regularly shown in television documentaries.
21. The Goebbels quote is from his diary entry of 29 July 1943; Albert Speer *Inside the Reich* pp. 283–4.
22. Max Hastings *Bomber Command* p. 257.
23. Henry Probert *Bomber Harris* p. 221.
24. *The Dam Busters*, producer Robert Clark, director Michael Anderson, starring Michael Redgrave as Barnes Wallis and Richard Todd as Guy Gibson, Associated British Picture Coproration, 1954. The film is based on a book by Paul Brickhill.
25. Albert Speer *Inside the Reich* pp. 280–1.

26. Martin Gilbert *Auschwitz and the Allies* p. 270.
27. Martin Gilbert *Auschwitz and the Allies* p. 285. For the recent debate about whether the Allies should have bombed Auschwitz, see William D. Rubinstein *The Myth of Rescue: Why the Democracies Could Not Have Saved More Jews from the Nazis* and Michael J. Neufeld and Michael Berenbaum (eds) *The Bombing of Auschwitz: Should the Allies Have Attempted It?* For the aerial photographs themselves and a summary of these issues, see the TV documentary *Auschwitz: The Forgotten Evidence*, producer Taylor Downing, director Lucy Carter, Flashback Television, 2005.
28. Max Hastings *Bomber Command* p. 244.
29. Winston Churchill *The Second World War Vol. IV* pp. 466 and 468.
30. See David Reynolds *In Command of History* pp. 320–4 and 396–8.
31. Martin Gilbert *Winston S. Churchill Vol. VII: Road to Victory* pp. 1160–1.
32. Martin Gilbert *Winston S. Churchill Vol. VII: Road to Victory* p. 1257.

Chapter 9 – Overlord

1. Ronald Lewin *Churchill as Warlord* p. 190
2. Martin Gilbert *Winston S. Churchill Vol. VII: Road to Victory* pp. 379–80.
3. Winston Churchill *The Second World War Vol. IV* p. 724.
4. Max Hastings *Overlord* p. 23. See also Alex Danchev and Daniel Todman (eds) *Field Marshal Lord Alanbrooke's War Diaries* p. 527.
5. Winston Churchill *The Second World War Vol. IV* pp. 729 and 741.
6. Winston Churchill *The Second World War Vol. V* p. 76.
7. Alex Danchev and Daniel Todman (eds) *Field Marshal Lord Alanbrooke's War Diaries* pp. 441–2.
8. Martin Gilbert *Winston S. Churchill Vol. VII: Road to Victory* p. 480 and see also p. 445.
9. Winston Churchill *The Second World War Vol. V* p. 123.
10. Ronald Lewin *Churchill as Warlord* p. 219.
11. Winston Churchill *The Second World War Vol. V* p. 128.
12. Martin Gilbert *Winston S. Churchill Vol. VII: Road to Victory* p. 497.
13. Martin Gilbert *Winston S. Churchill Vol. VII: Road to Victory* pp. 522 and 524 and Ronald Lewin *Churchill as Warlord* p. 222.
14. Martin Gilbert *Winston S. Churchill Vol. VII: Road to Victory* pp. 539 and 543.
15. Quoted from the Stilwell Papers in Ronald Lewin *Churchill as Warlord*

p. 228. See also Alex Danchev and Daniel Todman (eds) *Field Marshal Lord Alanbrooke's War Diaries* p. 478. Brooke had wanted the conference to be over before it had begun, knowing how unpleasant it would be 'the most unpleasant we have had yet, and that is saying a good deal'. *Field Marshal Lord Alanbrooke's War Diaries* p. 475.

16. Martin Gilbert *Winston S. Churchill Vol. VII: Road to Victory* pp. 578–9.
17. Winston Churchill *The Second World War Vol. V* pp. 329–30.
18. Lord Moran *Churchill at War* p. 171.
19. Martin Gilbert *Winston S. Churchill Vol. VII: Road to Victory* p. 586 and Winston Churchill *The Second World War Vol. V* p. 338.
20. Winston Churchill *The Second World War Vol. V* p. 339 and Ronald Lewin *Churchill as Warlord* p. 231.
21. Winston Churchill *The Second World War Vol. V* p. 371.
22. Martin Gilbert *Winston S. Churchill Vol. VII: Road to Victory* p. 602 and Winston Churchill *The Second World War Vol. V* p. 372.
23. Winston Churchill *The Second World War Vol. V* pp. 372–87, Martin Gilbert *Winston S. Churchill Vol. VII: Road to Victory* pp. 604–28, Lord Moran *Churchill at War* pp. 181–91, Alex Danchev and Daniel Todman (eds) *Field Marshal Lord Alanbrooke's War Diaries* pp. 497–8, and Sir John Martin *Downing Street: The War Years* pp. 124–33.
24. Martin Gilbert *Winston S. Churchill Vol. VII: Road to Victory* p. 771.
25. Winston Churchill *The Second World War Vol. II* p. 602 and John Colville *The Fringes of Power* p. 275.
26. Patrick Delaforce *Churchill's Secret Weapons* pp. 27–8.
27. Alex Danchev and Daniel Todman (eds) *Field Marshal Lord Alanbrooke's War Diaries* pp. 516–17.
28. The minute is reproduced as a facsimile with Churchill's handwritten notes in Winston Churchill *The Second World War Vol. V* opposite p. 78; see also J. Evans, E. Palmer and R. Walter (eds) *A Harbour Goes to War* pp. 5ff.
29. Hastings Ismay *Memoirs* p. 309.
30. Alex Danchev and Daniel Todman (eds) *Field Marshal Lord Alanbrooke's War Diaries* p. 554.
31. Hansard *Parliamentary Debates* House of Commons, 6 June 1944.
32. Martin Gilbert *Winston S. Churchill Vol. VII: Road to Victory* p. 802.
33. Winston Churchill *The Second World War Vol. VI* pp. 10–12 and Alex Danchev and Daniel Todman (eds) *Field Marshal Lord Alanbrooke's War Diaries* pp. 556–8.

Chapter 10 – Victory and Defeat

1. Alex Danchev and Daniel Todman (eds) *Field Marshal Lord Alanbrooke's War Diaries* p. 424.
2. Winston Churchill *The Second World War Vol. V* p. 206.
3. Dwight D. Eisenhower *Crusade in Europe* p. 260.
4. Alex Danchev and Daniel Todman (eds) *Field Marshal Lord Alanbrooke's War Diaries* pp. 521, 525 and 528.
5. Ronald Lewin *Churchill as Warlord* p. 238.
6. John Keegan (ed.) *Churchill's Generals* p. 99.
7. Alex Danchev and Daniel Todman (eds) *Field Marshal Lord Alanbrooke's War Diaries* p. 533.
8. Alex Danchev and Daniel Todman (eds) *Field Marshal Lord Alanbrooke's War Diaries* pp. 535 and 537.
9. Alex Danchev and Daniel Todman (eds) *Field Marshal Lord Alanbrooke's War Diaries* p. 544.
10. Captain Harry C. Butcher *Three Years with Eisenhower* p. 545; see also Ronald Lewin *Churchill as Warlord* p. 251 and Martin Gilbert *Winston S. Churchill Vol. VII: Road to Victory* p. 877.
11. Winston Churchill *The Second World War Vol. VI* p. 85 and Martin Gilbert *Winston S. Churchill Vol. VII: Road to Victory* p. 899.
12. Winston Churchill *The Second World War Vol. VI* p. 198. Churchill tells this story in the last volume of his memoir–history, published in 1954, after Stalin's death. Martin Gilbert notes that some of the official record of this discussion was later removed as it might 'seem most inappropriate for a record of this importance': see Martin Gilbert *Winston S. Churchill Vol. VII: Road to Victory* p. 992. The original piece of paper clearly showing Stalin's blue tick is reproduced in Jeremy Isaacs and Taylor Downing *Cold War* p. 12.
13. Winston Churchill *The Second World War Vol. VI* p. 205.
14. Martin Gilbert *Winston S. Churchill Vol. VII: Road to Victory* p. 1017.
15. Mary Soames *Clementine Churchill* p. 364.
16. Alex Danchev and Daniel Todman (eds) *Field Marshal Lord Alanbrooke's War Diaries* p. 673.
17. Alex Danchev and Daniel Todman (eds) *Field Marshal Lord Alanbrooke's War Diaries* pp. 676–7.
18. Martin Gilbert *Winston S. Churchill Vol. VII: Road to Victory* p. 1289.
19. Stephen Ambrose *Eisenhower and Berlin 1945: The Decision to Halt at the Elbe* pp. 53–65.

20. John Colville *The Fringes of Power* pp. 595–6.
21. Alex Danchev and Daniel Todman (eds) *Field Marshal Lord Alan-brooke's War Diaries* p. 687.
22. Elizabeth Nel (née Layton) *Mr Churchill's Secretary* p. 176.
23. Martin Gilbert *Winston S. Churchill Vol. VII: Road to Victory* p. 1347.
24. Winston Churchill *The Second World War Vol. VI* pp. 498–9.
25. Paul Addison *The Road to 1945* pp. 150–1.
26. Winston Churchill *The Second World War Vol. VI* p. 583.

Chapter 11 – Churchill's War

1. John Colville *The Fringes of Power* p. 509.
2. David Reynolds *In Command of History* pp. 28ff.
3. John Colville *The Churchillians* p. 143.
4. Robert Rhodes James *Churchill: A Study in Failure* is an early example, from 1970; Clive Ponting *Churchill* (1994) and Nigel Knight *Churchill: The Greatest Briton Unmasked* (2008) are later examples of this more hostile take on Churchill.
5. Speech given at Westminster Hall, London, 30 November 1954, on his eightieth birthday, quoted in David Cannadine (ed.) *Winston Churchill: Blood, Toil, Tears and Sweat; The Great Speeches* pp. 334–7.
6. Asa Briggs *The War of Words* pp. 9–10.
7. R.V. Jones *Most Secret War* p. 107.
8. Elizabeth Nel (née Layton) *Mr Churchill's Secretary* p. 58 and in interview with Flashback Television in 2006.
9. Nigel Knight *Churchill: The Greatest Briton Unmasked passim*. One of the great British military strategists, Basil Liddell Hart, also criticised Churchill's strategic vision in A.J.P. Taylor *et al. Churchill: Four Faces and the Man* pp. 155–202.
10. David Reynolds '1940: The Worst and Finest Hour' in Robert Blake and Wm. Roger Louis (eds) *Churchill* p. 249.
11. Martin Gilbert *Winston S. Churchill Vol. III: 1914–1916* p. 693.
12. Ruth Ive *The Woman Who Censored Churchill* p. 75 and in an interview for *Churchill and the President*, producer Taylor Downing, director Patrick King, Flashback Television, 1999.

Bibliography

Books by Winston Churchill

The World Crisis 1911–1918 First published in 5 vols, Odhams, London 1923–31; republished in 2 vols, Odhams, London, 1939
My Early Life Thornton Butterworth, London, 1930; republished Eland, London, 2000
Thoughts and Adventures Thornton Butterworth, London, 1932
Great Contemporaries Thornton Butterworth, London, 1937
Marlborough: His Life and Times 4 vols, Harrap & Co, London, 1933–8
The Second World War 6 vols, Cassell, London, 1948–54

Official biography

Randolph Churchill *Winston S. Churchill Vol. I: Youth 1874–1900* Heinemann, London, 1966
Randolph Churchill *Winston S. Churchill Vol. II: Young Statesman 1901–1914* Heinemann, London, 1967
Martin Gilbert *Winston S. Churchill Vol. III: 1914–1916* Heinemann, London, 1971
Martin Gilbert *Winston S. Churchill Vol. IV: 1916–1922* Heinemann, London, 1975
Martin Gilbert *Winston S. Churchill Vol. V: 1922–1939* Heinemann, London, 1976

Martin Gilbert *Winston S. Churchill Vol. VI: Finest Hour 1939–1941* Heinemann, London, 1983

Martin Gilbert *Winston S. Churchill Vol. VII: Road to Victory 1941–1945* Heinemann, London, 1986

Complementing these biographies are several volumes of documents edited by Martin Gilbert called *Companion* volumes, the most relevant of which for this study are:

The Churchill War Papers Vol. I: At the Admiralty September 1939 to May 1940 Heinemann, London, 1993

The Churchill War Papers Vol. II: Never Surrender May 1940 to December 1940 Heinemann, London, 1994

The Churchill War Papers Vol. III: The Ever Widening War 1941 Heinemann, London, 2000

Wartime memoirs and accounts by those who worked with or knew Churchill:

Field Marshal Lord Alanbrooke *War Diaries 1939–1945* ed. by Alex Danchev and Daniel Todman, Weidenfeld and Nicolson, London, 2001

Maurice Ashley *Churchill as Historian* Charles Scribner's Sons, New York, 1968

P.M.S. Blackett *Studies of War* Oliver & Boyd, Edinburgh, 1962

Violet Bonham Carter *Winston Churchill as I Knew Him* Eyre and Spottiswoode, London, 1965

Joan Bright Astley *The Inner Circle: A View of War at the Top* First published Hutchinson, London, 1971; republished The Memoir Club, Stanhope, 2007

Harry C. Butcher *My Three Years with Eisenhower* Simon & Schuster, New York, 1946

John Colville *The Churchillians* Weidenfeld and Nicolson, London, 1981

John Colville *The Fringes of Power – Downing Street Diaries 1939–45* Hodder and Stoughton, London, 1985

Andrew Cunningham *Admiral A.B. Cunningham: A Sailor's Odyssey* Hutchinson, London, 1951

Dwight D. Eisenhower *Crusade in Europe* Doubleday, New York, 1948

Sir Arthur Harris *Bomber Offensive* First published Collins, London, 1947; republished Greenhill, London, 1990

Hastings Ismay *The Memoirs of Lords Ismay* Heinemann, London, 1960

Ruth Ive *The Woman Who Censored Churchill* The History Press, Stroud, 2008

R.V. Jones *Most Secret War* Hamish Hamilton, London, 1978

Harold Macmillan *The Blast of War 1939–45,* Macmillan, London, 1977

Harold Macmillan *War Diaries: Politics and War in the Mediterranean, January 1943 to May 1945* Macmillan, London, 1984

Ivan Maisky *Memoirs of a Soviet Ambassador: The War 1939–43* trans. by Andrew Rothstein, Hutchinson, London, 1967

James Marshall-Cornwall *Rumours of War* Martin Secker & Warburg, London, 1984

Sir John Martin *Downing Street: The War Years* Bloomsbury, London, 1991

Field Marshal Viscount Montgomery of Alamein *Normandy to the Baltic* Hutchinson, London, 1947

Joanna Moody *From Churchill's War Rooms: Letters of a Secretary 1943–45* Tempus, Stroud, 2007

Lord Moran *Churchill at War 1940–45* First published Constable, London, 1966; republished Robinson, London, 2002

Elizabeth Nel (née Layton) *Mr Churchill's Secretary* First published Hodder and Stoughton, London, 1958; republished as *Winston Churchill by His Personal Secretary* iUniverse, New York, 2007

Lord Normanbrook (and others) *Action This Day: Working with Churchill* Macmillan, London, 1968

Mary Soames *Clementine Churchill* Cassell, London, 1979

Mary Soames (ed.) *Speaking for Themselves: The Personal Letters of Winston and Clementine Churchill* Doubleday, London, 1998

Robert E. Sherwood *The White House Papers of Harry L. Hopkins* Eyre & Spottiswoode, London, 1948

Albert Speer *Inside the Reich* Weidenfeld and Nicolson, London, 1970

Solly Zuckerman *From Apes to Warlords* Hamish Hamilton, London, 1978

Secondary accounts

Paul Addison *The Road to 1945: British Politics and the Second World War* Jonathan Cape, London, 1975

Paul Addison *Churchill on the Home Front* Jonathan Cape, London, 1992

Stephen Ambrose *Eisenhower and Berlin 1945: The Decision to Halt at the Elbe* Norton & Co, New York, 1967

Stephen Ambrose *D-Day* Simon & Schuster, New York, 1994

Constance Babington Smith *Evidence in Camera: The Story of Photographic Intelligence in the Second World War* First published Chatto & Windus, London, 1957; republished Sutton, Stroud, 2004

Joseph Balkoski *Beyond the Beachhead: The 29th Infantry Division in Normandy* Stackpole Books, Mechanicsburg, 1989

Correlli Barnett *The Audit of War* Macmillan, London, 1986

Geoffrey Best *Churchill: A Study in Greatness* Penguin, London, 2002

Geoffrey Best *Churchill and War* Hambledon, London, 2005

Lord Birkenhead *The Prof in Two Worlds: The Official Life of Professor F.A. Lindemann Viscount Cherwell* Collins, London, 1961

Robert Blake and Wm. Roger Louis (eds) *Churchill* Oxford University Press, 1993

Piers Brendon *Winston Churchill: A Brief Life* Pimlico, London, 2001

Asa Briggs *The History of Broadcasting in the United Kingdom Vol. III: The War of Words* 2nd revised edition, Oxford University Press, Oxford, 1995

Richard Broad and Suzie Fleming (eds) *Nella Last's War: A Mother's Diary 1939–45* Falling Wall Press, Bristol, 1981

William F. Buckingham *Arnhem 1944* Tempus, Stroud, 2002

Robert Buderi *The Invention that Changed the World: How a Small Group of Radar Pioneers Won the Second World War* Simon & Schuster, New York, 1996

Angus Calder *The People's War: Britain 1939–45* Jonathan Cape, London, 1969

Peter Calvocoressi and Guy Wint *Total War* Penguin, London, 1972

David Cannadine *The Decline and Fall of the British Aristocracy* Yale University Press, London, 1990

David Cannadine (ed.) *Winston Churchill: Blood, Toil, Tears and Sweat; The Great Speeches* Penguin Classics, London, 2007

Ronald W. Clark *The Rise of the Boffins* Phoenix House, London, 1962

Ronald W. Clark *Tizard* Methuen, London, 1965

John Costello *The Pacific War 1941–1945* HarperCollins, New York, 1982

Len Deighton and Max Hastings *Battle of Britain* Jonathan Cape, London, 1980

Patrick Delaforce *Churchill's Secret Weapons: The Story of Hobart's Funnies* Pen & Sword, Barnsley, 2006

Bob Doe *Fighter Pilot* CCB Associates, Selsdon, 1999

Taylor Downing and Andrew Johnston *Battle Stations: Decisive Weapons of the Second World War* Pen & Sword, Barnsley, 2000

J. Evans, E. Palmer and R. Walter (eds) *A Harbour Goes to War: The Story of Mulberry and the Men Who Made It Happen* Brook House Publishing, Wigtownshire, 2000

David Fraser *Alanbrooke* HarperCollins, London, 1982

Juliet Gardiner *Wartime Britain 1939–45* Headline Books, London, 2004

Martin Gilbert *Churchill: A Life* Heinemann, London, 1991

Martin Gilbert *Auschwitz and the Allies* Pimlico, London, 2001

Ian Grant *Cameramen at War* Patrick Stephens, Cambridge, 1980

Max Hastings *Bomber Command* Penguin, London, 1999

Max Hastings *Overlord* Pan Macmillan, London, 1999

Richard Havers *Here is the News: The BBC and the Second World War* Sutton Publishing, Stroud, 2007

F.H. Hinsley and Alan Stripp (eds) *Code Breakers: The Inside Story of Bletchley Park* Oxford University Press, Oxford, 1993

Richard Holmes *In the Footsteps of Churchill* BBC Books, London, 2006

Richard Holmes *The World at War: The Landmark Oral History* Ebury, London, 2007

Alistair Horne with David Montgomery *The Lonely Leader: Monty 1944–1945* Macmillan, London, 1994

Jeremy Isaacs and Taylor Downing *Cold War* Transworld, London, 1998 and republished by Little, Brown, London, 2008

Brian Johnson *The Secret War* BBC Books, London, 1978

David Kahn *Seizing the Enigma* Arrow, London, 1996

John Keegan *Six Armies in Normandy: From D-Day to the Liberation of Paris* Jonathan Cape, London, 1982

John Keegan *Churchill* Weidenfeld & Nicolson, London, 2002

John Keegan (ed.) *Churchill's Generals* Abacus, London, 1999

Warren Kimball *Forged in War: Roosevelt, Churchill and the Second World War* William Morrow and Company, New York, 1997

Nigel Knight *Churchill: The Greatest Briton Unmasked* David & Charles, Newton Abbot, 2008

Ronald Lewin *Churchill as Warlord* Batsford, London, 1973

Ronald Lewin *Ultra Goes to War: The Secret Story* Hutchinson, London, 1978

Adrian R. Lewis *Omaha Beach: A Flawed Victory* Tempus, Stroud, 2004

John Lukacs *Five Days in London May 1940* Yale University Press, London, 1999

Leo McKinstry *Spitfire: Portrait of a Legend* John Murray, London, 2007

Michael J. Neufeld and Michael Berenbaum (eds) *The Bombing of Auschwitz: Should the Allies Have Attempted It?* University Press of Kansas, Lawrence, 2003

Richard Overy *Russia's War* Allen Lane, London, 1998

Richard Overy *The Battle of Britain* Penguin, London, 2000

Clive Ponting *Churchill* Sinclair-Stevenson, London, 1994

Henry Probert *Bomber Harris: His Life and Times* Greenhill, London, 2003

Nicholas Rankin *Churchill's Wizards: The British Genius for Deception 1914–1945* Faber and Faber, London, 2008

David Reynolds *In Command of History: Churchill Fighting and Writing the Second World War* Allen Lane, London, 2004

Robert Rhodes-James *Churchill: A Study in Failure 1900–39* Weidenfeld and Nicolson, London, 1970

Andrew Roberts *Hitler and Churchill: Secrets of Leadership* Phoenix, London, 2003

Andrew Roberts *Masters and Commanders: How Roosevelt, Churchill, Marshall and Alanbrooke Won the War in the West* Allen Lane, London, 2008

Andrew Roberts *The Storm of War: A New History of the Second World War* Allen Lane, London, 2009

Stephen Roskill *Churchill and the Admirals* First published William Collins, London, 1977; republished as a Pen & Sword Military Classic, Barnsley, 2004

William D. Rubinstein *The Myth of Rescue: Why the Democracies Could Not Have Saved More Jews from the Nazis* Routledge, London, 1997

Celia Sandys *Churchill Wanted Dead or Alive* HarperCollins, London, 1999

L.A. Sawyer and W.H. Mitchell *The Liberty Ships* Lloyds of London, London, 1985

Tom Shachtman *Laboratory Warriors: How Allied Science and Technology Tipped the Balance in World War Two* HarperCollins, New York, 2003

Gary Sheffield and John Bourne (eds) *Douglas Haig: War Diaries and Letters* Phoenix, London, 2006

Michael Smith *Station X: The Codebreakers of Bletchley Park* Channel Four Books, London, 1998

David Stafford *Roosevelt and Churchill: Men of Secrets* Little, Brown, London, 1999

David Stafford *Churchill and Secret Service* Abacus, London, 2000

A.J.P. Taylor *English History 1914–1945* Oxford University Press, Oxford, 1965

A.J.P. Taylor, R.R. James, J.H. Plumb, B.L. Hart and A. Storr *Churchill: Four Faces and the Man* Allen Lane, London, 1969

Sir Charles Webster and Noble Frankland *The Strategic Air Offensive against Germany 1939–1945* 4 vols, HMSO, London, 1961

John Wheeler-Bennett *King George VI: His Life and Reign* Macmillan, London, 1958

Thomas Wilson *Churchill and the Prof* Cassell, London, 1995

Solly Zuckerman *Scientists and War: The Impact of Science on Military and Civil Affairs* Hamish Hamilton, London, 1966

Acknowledgements

I have been lucky enough to have produced many documentaries at Flashback Television relating to several of the subject areas covered in this book. And I have been privileged to interview some of the individuals who feature in the book or who knew those who feature. So the ideas here have been mulling over in my mind for many years after discussions and debate with a large number of people whom I would like to thank. First of all, there is my business and creative partner David Edgar, who deserves special thanks for coming up with the title. Then there are many other fellow travellers with whom I have had the pleasure of working on television documentaries in this area. They include Andrew Johnston, Patrick King, Colin Barratt, Chris Warren, David Caldwell-Evans, Steve Baker, Paul Nelson, Jobim Sampson, James Barker, Lucy Carter and Dunja Noack. I need to thank all of them. Also, working with Sir Jeremy Isaacs has always been pleasurable and rewarding. I have benefited from discussions with several professional historians, including Professors David Cannadine, Richard Overy, Richard Holmes and Gary Sheffield. Phil Reed, director of the Imperial War Museum's Cabinet War Rooms and Churchill Museum, has always been wonderfully helpful and open.

I should like to thank Tim Whiting, Iain Hunt and Philip Parr at Little, Brown for their continuing support and encouragement, and Linda Silverman for the photo research; and the staff of the London Library for their efficient support. Extracts from the writings of Winston Churchill are reproduced with permission of Curtis Brown Ltd, London, on behalf of The Estate of Winston Churchill, copyright © Winston S. Churchill. Finally, my thanks, as always, go to Anne for her patience and her support.

Index

COLD WAR

Jeremy Isaacs and
Taylor Downing

Cold War describes the events of the forty-five year confrontation
between two great powers that has defined the modern world.
It is a story of the constant threat of nuclear destruction and
of crises and conflicts on a global scale: of the Berlin Blockade
and the Cuban Missile Crisis, of savage wars, of tanks in the
streets of Warsaw, Berlin and Budapest; of spies, student riots
and encounters in space. It is also a story of shifting economic
fortunes and dramatic social change, of oil price rises,
blue jeans, rock 'n' roll and inner-city violence.

Meticulously researched and richly detailed, *Cold War* draws on
historical archives and individual testimony – of statesmen and
ordinary people alike – to give a unique and gripping insight
into the events of this unprecedented period.

ABACUS
978-0-349-12080-5